FORESTRY COMMISSION BULLETIN 102

Forest Fencing

H. W. Pepper
Forestry Commission

LONDON: HMSO

© *Crown copyright 1992*
  First published 1972 (as Forest Record 80)
  Third edition 1992

ISBN 0 11 710304 7
ODC 451.4 : 451 : 413.5

KEYWORDS: Fencing, Forestry, Protection, Wildlife

Enquiries relating to this publication
should be addressed to:
The Technical Publications Officer,
Forestry Commission, Forest Research Station,
Alice Holt Lodge, Wrecclesham,
Farnham, Surrey, GU10 4LH.

Front cover: Deer fencing. *(40277)*
Inset: Straining the wire. *(40279)*

Contents

	Page
List of illustrations	v
List of tables	vi
Summary	vii
Résumé	viii
Zusammenfassung	ix
Introduction	1
Fencing principles and specifications	2
Principles	2
When to fence	2
Where to fence	2
What to use	3
Specifications	5
Rabbit fences	6
Stock fences	7
Mines and quarries fence	7
Deer fence – light specification	8
Deer fence – heavy specification	9
Fence components and associated tools	11
Metalwork	11
Wire and associated tools	11
Wire fixings	14
Wire netting and associated tools	16
Woodwork	21
Durability, sizes and associated tools	21
Fencing belt	23
Fence construction and maintenance	23
Construction	23
Spacing and erecting the posts	24
Strutting the posts	26
Contour and turning posts	27
Straining the wire	28
Joining the wire	30
Spacing the stakes	30

Fixing wire to stakes	32
Erecting the netting	32
Sequence of operations	33
A method of piece-work payment	33
Safety	35
Maintenance	35
Acknowledgements	36
References	37
Appendix 1 Manufacturers' addresses, products and specifications	39
Appendix 2 Badger gates	41

List of illustrations

Figure	Subject	Page
1	Rabbit fence specification	6
2	An immature rabbit going through 36 mm mesh	7
3	Stock fence specification	8
4	Mines and quarries fence specification	8
5	Deer fence – light specification for roe	9
6	Deer fence – heavy specification for fallow, red and sika	10
7	Watergate specification	11
8	Fencing terms	12
9	Graphs showing behaviour of steel wires when tensioned	13
10	Fencing tools	14
11	Wire dispensers with wire	15
12	Types of barbed wire	15
13	Ratchet winder	16
14	Preformed fixings	16
15	Wire connectors	16
16	'Double-six' knot and 'Double-loop' knot	17
17	Woven wire mesh netting joints	17
18	Netting sizes and patterns	18
19	Sharp's straining bar	19
20	Wire-ring straining bar	19
21	Clamp straining bar	20
22	The 'Tornado' professional clamp	20
23	Gerrard ring gun	21
24	Shape of post hole	23
25	Fencing tools	23
26	Fencing belt with tools	24
27	Sketch-map for estimating quantities of materials	25
28	Cross-member on post	25
29	Relative positions of cross-members, struts, posts and line wires at changes of direction of fence-line	26
30	End post assembly	26
31	Strut notched to round thrust-plate	27
32	The method of stapling a wire to a post	27
33	Straining the retaining wire and securing with a fence connector	28
34	Straining the retaining wire and securing with a wirelok	29
35	Fence connector terminating line wire	29
36	Points on a line wire at which fence connector is attached	30
37	Fence connectors joining line wires	30
38	Spacing stakes	31

39	Wire-netting dispenser	32
40	Joining woven or welded wire mesh netting	33
41	Straining woven or welded wire mesh netting	34
42	Removing strain from fence connector	36

All illustrations are drawn from Forestry Commission sources.

List of tables

Table	Subject	
1	Woodwork sizes – rabbit and stock	7
2	Woodwork sizes – sheep and cattle	7
3	Woodwork sizes – roe deer	8
4	Woodwork sizes – red, sika or fallow deer	9
5	Wire specifications	12
6	Preformed fixings – sizes and uses	17

Forest Fencing

Summary

Fencing is a necessary but expensive forest management operation. While it is possible to erect a fence that is completely effective against any animal, this is usually too costly. Any forest fence is a compromise between expense and effectiveness.

The introduction of spring steel wire by the Forestry Commission into the construction of forest fences in 1969 substantially reduced costs compared with traditional forms of fencing using mild steel wire. The main advantage of spring steel wire is that once it is strained it will remain taut. This allows stakes and straining posts to be widely spaced, so requiring less material and labour without reducing the effectiveness of the fence. Small economies have been made by using spring steel wire for multi-strand and dropper fences but the most suitable application and the greatest economies are in the use of this wire to support wire netting. Further savings have been made by introducing improved methods of working and labour-saving tools.

Improvements to the design and construction of fences are constantly under review. In the 23 years since 1969 new and improved materials, tools and working practices have been introduced.

Any savings obtained can be wasted if the initial planning of the fence has not been thorough. The specification of the materials to be used must be consistent with the period and the purpose for which the fence is required. The amount of material required can be reduced and the problems of negotiating natural obstacles can be avoided by careful siting.

La Clôture des Forêts

Résumé

La pose de clôtures est une opération qui s'avère indispensable mais très coûteuse dans le cadre de l'exploitation des forêts. En effet, il est tout à fait possible de poser une clôture de protection contre tous les animaux, mais le coût correspondant est généralement beaucoup trop élevé. Donc le choix d'une clôture résulte d'un compromis entre le coût impliqué et l'efficacité de la clôture choisie.

En introduisant l'utilisation de fils d'acier à haute tension dans la construction des clôtures de forêts en 1969, le service de Eaux et Forêts (Forestry Commission) a permis de réduire considérablement les coûts par rapport à certains types de clôtures traditionnelles utilisant des fils en acier doux. L'intérêt des fils d'acier à haute tension est qu'une fois étirés, ils restent tendus. Cela permet d'espacer davantage les piquets et les poteaux tendeurs, tout en nécessitant moins de matériaux et moins de main-d'oeuvre, l'efficacité de la clôture restant identique. L'utilisation de fils d'acier à haute tension dans les clôtures comportant de nombreux torons et piquets d'espacement a permis de réaliser quelques économies, mais elle s'avère surtout intéressante lorsque les fils d'acier à haute tension servent de supports à l'intérieur des grillages. D'autres économies ont été réalisées grâce à l'introduction de nouvelles méthodes de travail et d'outillages qui permettent de réduire la main-d'oeuvre nécessaire.

Les progrès possibles en matière de conception et construction de clôtures font l'objet de recherches continuelles. Au cours des 23 ans qui ont suivi l'année 1969, de nouvelles méthodes de travail et de nouveaux matériaux et outils plus performants ont été introduits.

Quoi qu'il en soit, les économies réalisées peuvent s'avérer inutiles si la pose de la clôture n'a pas été planifiée dès le début avec minutie. Les propriétés des matériaux utilisés doivent satisfaire plusieurs critères à savoir la durée d'utilisation de la clôture et le rôle de celle-ci. La quantité de materiaux nécessaires peut être réduite et le problème posé par les obstacles naturels peut être évité grâce à un positionnement judicieux de la clôture.

Forsteinzäunung

Zusammenfassung

Einzäunung ist ein wichtiger aber teurer Teil des Forstbetriebs. Obwohl es möglich ist, einen Zaun zu errichten, der gegen alle Tiere wirksam ist, sind die Kosten dafür normalerweise zu hoch. Jede Forsteinzäunung stellt einen Kompromiß zwischen Kosten und Wirksamkeit dar.

Die Einführung von Federstahldraht in die Konstruktion von Forstzäunen durch die Forstverwaltung im Jahr 1969 reduzierte die Kosten im Vergleich zu traditionellen Einzäunungsarten sehr stark, bei denen weicher unlegierter Stahldraht verwendet wird. Der größte Vorteil von Federstahldraht ist, daß er nach dem Ziehen straff bleibt. Somit können Pfosten und Spannpfosten in großen Abständen angebracht werden, wodurch weniger Material und Arbeitszeit benötigt wird, ohne die Wirksamkeit des Zauns zu vermindern. Geringe Einsparungen wurden durch den Gebrauch von Federstahldraht bei Mehrstrang- und Hängezäunen erzielt, aber dieser Draht ist am besten als Stütze für Drahtgeflechte geeignet. Dadurch entstehen die größten Einsparungen. Weitere Einsparungen wurden durch die Einführung von verbesserten Arbeitsmethoden und arbeitssparenden Werkzeugen erzielt.

Verbesserungen in Design und Konstruktion von Zäunen werden laufend überprüft. In den 23 Jahren seit 1969 wurden neue und verbesserte Materialien, Werkzeuge und Arbeitsmethoden eingeführt.

Erzielte Einsparungen können jedoch wieder verschwendet werden, wenn die Einzäunung nicht gründlich geplant wurde. Die technischen Angaben der zu verwendenden Materialien müssen für den Zeitraum und Zweck, für die der Zaun benötigt wird, geeignet sein. Die benötigten Materialmengen können verringert und Probleme, die durch ein Ausweichen von natürlichen Hindernissen entstehen, vermieden werden, wenn das Aufstellen des Zaunes gründlich im voraus geplant wird.

Forest Fencing

H. W. Pepper, *Wildlife and Conservation Research Branch, Forestry Commission*

Introduction

Post and wire fences are to be found in every county of the British Isles where there are forestry and farming interests. Most of the techniques used in the erection of these fences are traditional: methods have been handed down from times when labour and materials were much cheaper than they are today. These fences were made up from a variety of woodwork ranging from cleft or squared oak to rough unsaleable softwood in conjunction with mild steel wire. Temperature changes cause this wire to stretch and slacken, so stakes have to be spaced three to five metres apart to achieve the desired strength and rigidity.

Fencing is an efficient method of protecting forest and farm crops from domestic stock and wild animals. It is also an extremely expensive operation. The high cost of fencing and the increasing demand for reduced establishment and protection costs in forestry made it imperative to investigate ways of reducing fencing costs.

During the course of the investigations by the Forestry Commission 2.65 mm diameter spring steel wire was introduced to replace mild steel wire. The wire referred to throughout this document as spring steel wire is grade 6 steel wire of BS 5216 (British Standards Institution, 1975). Although of a smaller gauge than the 4.00 mm and 3.15 mm mild steel wires previously used, spring steel wire has superior properties. It is much stronger and because it has some elasticity it will retain its tension since, unlike mild steel wire, it will not stretch and slacken under normal tension. The cost per metre of spring steel wire is no greater than that of mild steel wire.

The stability of spring steel wire, under tension, allows the distances between straining posts and stakes to be increased without reducing the rigidity and strength of the fence. For example, distances of up to 1000 m between straining posts have been attained over ideal conditions of gently undulating topography, and on flat terrain a 14 m interval between stakes has been found satisfactory. The reduced amount of woodwork and the consequent saving in labour contributes a major part of the saving in cost of spring steel fencing over traditional fencing with mild steel wire.

A minor disadvantage of spring steel wire is that it is more difficult to handle than mild steel wire. This is overcome by the use of a few special tools.

At the same time as the spring steel wire was being tested the whole technique of fence erection was studied. Each operation was carefully examined to see if it could be improved and the work content reduced. In addition, the specification of the supporting woodwork and of the netting suitable for different types of fence was investigated. As a result a new form of netting fence was evolved, together with a new erection system, to replace traditional fence designs and erection methods. The system, introduced in 1969, included the use of spring steel wire with its specialised tools together with improved methods of post, stake and netting erection using a two-man team with a standardised set of tools. Since then, further improvements have come as a result of new techniques in netting manufacture and the introduction of mesh sizes more suited to forest fences which are principally required to exclude wild animals as opposed to enclosing domestic stock. All fence components can now be obtained in a quality that complies with one or more of the British Standards Institution's specifications.

Recent developments in electric fencing are being assessed to determine its efficacy against wild animals, particularly around restock areas and farm woodlands. It does appear to be less suitable for forest protection than for agricultural purposes where fences are more accessible to maintain and the movable temporary fence is often a necessary requirement.

Improving the system is a continuing process and at the time of writing new designs of wire and wire mesh strainers are being evaluated together with a small compact device for joining line wire. The minimum height requirement for a roe deer fence is also being investigated.

Fencing principles and specifications

Principles

When to fence

'Is the fence really necessary?' This is the vital question that must be answered before fencing any area. In some cases a fence may be obligatory because of some agreement in a lease or statute. If marker posts are insufficient, a fence may also be required to mark a boundary. In these cases the answer to the question is a straightforward 'yes'. Arriving at an answer may not be so simple when the object of a fence is purely to protect forest trees.

An estimate of the amount of damage animals may do to the plantation at risk will have to be made to determine what the result will be if the area is not fenced. As a decision will usually need to be made before the crop is planted, the estimate will often have to be based on assessments of damage (Melville, Tee and Rennolls, 1983) in similarly situated unfenced crops. Perhaps at best satisfactory crop establishment would be delayed by a year or two because of the necessity to beat up, or at worst the crop could not be established at all. The probable amount of damage that one or more species of animals would be prevented from doing to a plantation must be compared with the cost of the fence. In other words, can the cost of the fence be compensated for from the extra value obtained from the final crop?

Animal damage may not in itself be significant but it may be the one controllable factor that could turn an uneconomic crop into an economic one. On the other hand fencing an area may be justified when related to the expected cash returns but it may not necessarily be the cheapest method of protecting the crop. Control of the animal or even a different choice of tree species may be alternatives which should always be considered. It should also be remembered that control of the animal population may be obligatory as in the case of the rabbit, or necessary as in the case of deer even if fencing is used to protect the crop. Individual tree protection in the form of guards (Pepper, Rowe and Tee, 1985), treeshelters (Potter, 1991) or chemical repellants (Pepper, 1978) can compete with fencing in effectiveness. The choice between fencing and individual tree protection is governed by the perimeter length of the plantation to be protected and the number of trees per hectare planted. In general on areas of less than 4 ha individual tree protection is more likely to be used and areas in excess of 4 ha fencing will be used (Ratcliffe and Pepper, 1987).

Where to fence

The location of a fence can influence the cost of fence erection and careful planning is necessary if the saving in cost obtained by using the most cost effective fence specification is not to be lost.

Cost, although a major consideration, is not the only factor to be taken into account when planning the location of the fence line. When new fences, and particularly boundary fences, are being planned all interested parties must be consulted so that the proposed fence line is acceptable to all. The line a fence may be required to take is often rigidly defined by law, or the geography of the area, allowing for little subsequent variation. Where this is not so, it may be possible to make worthwhile savings by straightening out the line to eliminate one or more corner posts even at the expense of excluding some land. A fence should not be used simply to mark an irregular boundary where there is an option to straighten the line.

The ease of digging-in and firming straining posts should also be considered, and waterlogged areas and shallow soils over rock avoided. Where possible, avoid fencing over excessively undulating ground where it may be difficult to prevent the fence lifting off the ground. By choosing a relatively level line fewer stakes are required and 'filling-in' (Figure 8) can be avoided. The extent and location of routine forest operations such as planting and harvesting may have an influence on the final choice of the fence line, as may landscaping and amenity requirements.

The length of fence in relation to the area to be enclosed must be taken into account. The aim should be to fence a given area with the minimum practical length of fence. It is an accepted fact that the shortest line required to encompass a given area would be in the form of a circle. This is for obvious reasons not practical and the next best shape to give the lowest ratio of length of fence to area enclosed is a square. It is also true that the larger the area fenced the lower the cost per hectare, but very large areas are often unmanageable if animals break in. The cost per hectare of fencing decreases considerably as the area increases to 40 ha. With areas in excess of 40 ha the decrease is negligible. The ideal fence therefore is one that encloses an area which is as near square as possible and is 40 ha or more in size.

The habits and movements of the animals to be excluded may cause management problems if not considered. Red deer on open-range for instance may put pressure on fences, especially when these follow the contours and prevent deer gaining access to sheltered valleys in inclement weather. This can be remedied by creating down-falls to allow access up and down the hill between fenced blocks.

Climatic conditions may also need to be considered. For example, in a valley that is subjected to regular drifting snow, a fence placed some yards up the hillside rather than in the valley bottom may miss the snow drifts. However, care should be taken when fencing along a steep slope that animals approaching from the higher side are not able to jump over the fence.

Every opportunity must be taken to gather and consider all the available relevant local knowledge before a fence line is agreed and the fence is erected. Relocation of a fence after completion is expensive.

What to use

The object of a fence must be clearly defined before any consideration is given to fence type and detailed specification. Therefore knowledge is required of:

1. the species and sometimes breed of animal or animals which are to be fenced either in or out and their capability to scale, burrow or just force their way through a fence;
2. pressure related to the number of animals on one side of the fence and their need to be on the other;
3. the length of time an effective fence is required; and
4. the maximum permitted level of financial expenditure.

Whatever the type of fence chosen, spring steel wire should be used in preference to mild steel wire. The overall reduction in cost of materials and labour alone justifies its use. There are no conceivable conditions where a mild steel fence could be erected but a spring steel one could not.

Spring steel wire of 2.65 mm diameter in conjunction with either woven, welded or hexagonal mesh wire netting is recommended as being the most effective and economical fence. Although it is possible to erect a fence entirely of spring steel wires it is unlikely to be cheaper and it is not recommended. The anchorage of the end posts has to be reinforced considerably to withstand six or more line wires all constantly at 4000 N. To maintain the wider stake spacing would mean the introduction of droppers. Metal droppers are expensive and wooden droppers are unreliable due to the tendency of the staples to pull out as the dropper seasons. The effectiveness of line wire and dropper fences against deer and sheep is dubious.

Sightings have been made of red deer hinds jumping between line wires 0.2 m apart and roe deer pushing through wires that are only 0.15 m apart. Similarly electric fencing, which is effectively multi-strand line wire fencing with pulsed electric power, has not proved to be as effective as wire and wire mesh despite the claims of some manufacturers. The use of 3.15 mm diameter spring steel line wires may be considered necessary when maximum fence life is required on areas with very high industrial or coastal atmospheric pollution. Galvanised netting and wire are recommended for all areas of Britain.

Wire and wire netting with a PVC bonded coat is no longer generally available for agricultural and forestry fencing even though it is widely available for high security fencing. Some mild steel wires and hexagonal mesh netting are *sleeved* with PVC. This process is inferior to bonding as it can easily be stripped off during erection or by vandals with penknives. Corrosion can also 'creep' under the PVC sleeve.

Different designs of electric fences and their efficacy in forest situations are being evaluated. The main reason for wanting to use electric fences is their apparent lower capital cost of installation when compared with netting fences. Unfortunately the running and maintenance costs may erode any initial saving (Roe and Tee, 1980).

The evidence from trials (Pepper et al., 1992) suggests that an electric fence against deer, and roe deer in particular, is not as effective a barrier as a line wire and wire mesh fence. Roe deer were deterred to some extent but browsing damage and deer activity within the fenced area were both unacceptably high.

A representative range of battery operated electric fence energisers was evaluated and none was found to comply with BS 6167 and some were considered unsafe. All performed differently when connected to fences of different load conditions. The prevailing load conditions are principally dependent on the fence specification, the fence length, the earthing quality and the level of insulation, i.e. presence of vegetation and weather conditions. Generally the greater the load present the lower energy of pulse and therefore a less severe shock.

The electrical impedance of the body of deer appears to differ considerably between species and therefore each species will experience different levels of shock sensation from the same fence. Red deer appear to be the most susceptible and roe deer the least. The shock experienced by a human, from the same fence, will be more severe than that experienced by deer.

More work is being and needs to be done to improve the design and performance of electric fences and electric fence equipment. It is therefore not possible, at the time of writing, to make any recommendations. However, if electric fencing can be made effective against wild animals in forest conditions it is likely that it will be principally line wire fencing using 2.65 mm diameter spring steel wire and therefore many of the basic principles described will be applicable.

The choice of netting to use on a fence will depend upon the animals to be excluded. Some suggestions of the most suitable height and mesh size of net for each type of fence are given on pages 5-11. In some cases the only suitable netting available at present is in the hexagonal mesh patterns, for example, rabbit netting. However, in general, welded wire mesh and woven wire mesh netting are cheaper and more robust than the comparable hexagonal mesh netting and are therefore preferred where there is a choice between the two types. When choosing a deer fence specification two widths of netting (Figures 5 and 6) are recommended in preference to single width 1.8 m or 2 m netting. The use of two nets allows flexibility of choice. Nets with different mesh sizes can be combined to tailor the fence to the specific needs of a particular situation. The width and weight (approximately 150 kg per roll) of the single width netting make it difficult to handle on all but the most accessible sites. It is therefore best suited to deer farm fencing for which it was originally designed. Heavy grade netting (pages 18-19) is recommended when a durable fence is required in areas of high atmospheric pollution.

The desirability of using treated or untreated timber is covered on pages 21-23. Wherever possible, round wood is preferred to squared or cleft wood for stakes because of its suitability for use with the 'Drivall' and its general ease of handling. The most suitable sizes of material for each type of fence are given on pages 6-11.

Gateways should be kept to a minimum and wherever possible gates should be hung on posts that are independent of the straining posts. Frequently used gateways should have gates that are easily opened and closed. Stiles or bridle gates must be provided where a fence crosses a right of way.

The final quality of the completed fence will depend on the quality of the materials and the workmanship of the labour. If the fencing is done by direct labour poor workmanship is avoided by providing the appropriate training and supervision. If contract labour is to be used it is necessary to specify in exact detail the standard expected and make regular inspections to ensure the specified standard is being attained, for example, that strainers are being put in to the required depth complete with cross-member and that stakes are being driven deep enough. It is prudent to employ a contractor who is registered under and complies with the quality assurance requirements of BS 5730. Under this standard the contractor must offer products and services that:

1. meet a well defined need, use or purpose;
2. satisfy customers' expectations;
3. comply with applicable standards and specifications;
4. comply with statutory (and other) requirements of society. These requirements include laws, statutes, rules and regulations, codes, environmental considerations, health and safety factors, and conservation of energy and materials;
5. are made available at competitive prices;
6. are provided at a cost which will yield a profit.

Similarly when ordering materials there must be complete control over the quality of the product purchased. All materials described in this Bulletin can be specified to comply with one or more British Standard Institution specifications. They should also be purchased from a manufacturer/supplier operating to BS 5750. If a contractor is to supply the materials as well as the labour, checks can be made (by inspecting labels, delivery notes, etc.) to verify that the materials comply with the relevant standards. It must be clearly stated if sub-standard items are acceptable for short-term or temporary fences.

There are no grants available specifically for forest fences. The Ministry of Agriculture, Fisheries and Food (MAFF) and the Scottish Office Agriculture and Fisheries Department (SOAFD) make investment grants for the provision of new or renewed permanent safety and protective fences on agricultural and horticultural holdings. Grants are paid either on standard costs – a fixed sum per metre on fences erected to a prescribed specification – or on actual costs of fences erected to approved standards. Evidence of payments made have to be provided to claim a grant, under the actual cost scheme. Boundary fences between forest and agriculture may qualify for a grant. The specifications shown on pages 6-11 are acceptable for actual cost grant aid. Standard cost aid is only paid on a restricted number of fence specifications and the rabbit fence (Figure 1) is the only one that is likely to be eligible. A grant application is made to either a MAFF Divisional Office or a SOAFD Regional Office.

Specifications

Each fence line is a unique combination of topography, climate, atmospheric pollution and external animal pressure. Spring steel line wire and netting fencing is a versatile system allowing the forest manager flexibility of choice over the type and pattern of net or combination of nets to be used, together with the sizes and spacing of woodwork that are most suited to the topography/exposure, the pressure from the animals being excluded and the prevailing weather conditions on his own area. A fence can be made more or less robust along particular lengths to cope with localised pressures. This section gives recommendations based on the minimum specifications necessary, all comply with the requirements of BS 1722 : Part 2. These specifications must be increased if the fences are to remain effective in the more difficult situations, for exam-

RABBIT FENCE

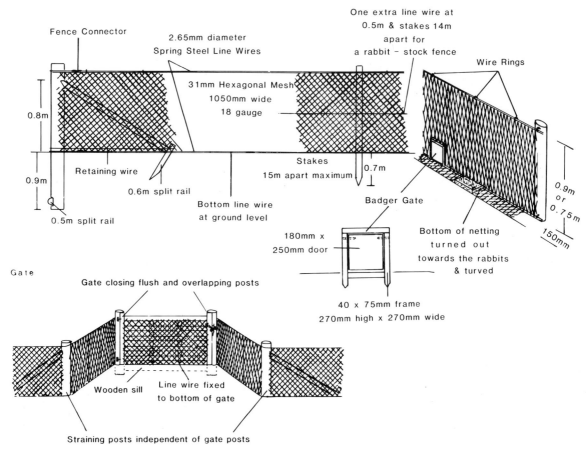

Not to scale. For descriptions of recommended netting and possible alternatives see Figure 18 page 18. Netting sizes and patterns.

Figure 1. Rabbit fence specification.

ple, in the uplands where conditions are particularly harsh.

Once a decision has been made on the fence specifications to be used, a detailed list of type, quantity and source of materials and tools required can be compiled using the specifications given on pages 6-11 in conjunction with sketch map (page 25) and Tables 1 to 6.

Rabbit fences

The basic specification for rabbit fencing is shown in Figure 1. The additional modifications required when it is necessary to exclude domestic stock in addition to rabbits are also detailed: these involve slightly larger and more robust posts and an extra line wire.

Raising the height above 0.9 m will not make the fence a more effective barrier to rabbits and it must not be allowed to drop below 0.675 m (McKillop, Pepper and Wilson, 1988). The maximum size of hexagonal mesh netting that can be specified to exclude rabbits is 31 mm. Some juvenile rabbits, capable of surviving without their mother, can pass through 36 mm mesh (Figure 2) (the next available size of hexagonal mesh). Rabbits are capable of gnawing through 19 gauge mild steel wire. Therefore netting of 18 gauge wire should be used for all permanent rabbit fences and 19 gauge netting should only be used for tempor-

Figure 2. An immature rabbit going through 36 mm mesh. *(A7982)*

ary (6 months to 3 years) fencing and then only if rabbit pressure is moderate (less than 10 rabbits per hectare). Netting 1200 mm wide dug in the ground 150 mm and lapped out horizontally 150 mm towards the rabbits is no more effective than using 1050 mm or 900 mm wide netting with 150 mm turned out on the surface and held down, as shown, with pegs and/or turves. Rabbit fencing is best done just before the growing season so that the vegetation grows through the lapped netting as quickly as possible to prevent rabbits finding a way underneath. Fencing over tree stumps and rock outcrops should be avoided as these can be difficult to make rabbit-proof.

The number of gates in a fence should be kept to a minimum because it is difficult to make and maintain rabbit-proof gates. They should be hung on reversed hinges and kept closed with a chain and padlock. Gate posts should be independent of the straining posts to prevent the slight but unavoidable movement of the straining posts altering the swing of the gate. A wooden sill is dug in between the gate posts to prevent burrowing under the gate. A concrete sill is preferable on metalled roads.

Badger gates (Rowe, 1976) must be provided wherever a fence crosses an established badger run (see Appendix 2).

The temptation to increase the recommended maximum distance between stakes must be resisted. A build up of vegetation on netting between over-widely spaced stakes will, with the addition of snow, weigh down the fence.

Stock fences

The basic specification for excluding sheep and cattle is detailed in Figure 3. When fencing against cattle many farmers require a barbed wire on the top of the fence in preference to the plain wire shown. However, where deer, especially fallow deer, are present the barbed wire should not be used because the deer on occasions drop a hind leg between the barbed wire and the netting and the leg can become irrevocably trapped, resulting in a slow and painful death.

Mines and quarries fence

This is a fixed specification (Figure 4) for secu-

Table 1. Woodwork sizes – rabbit and stock

Rabbit	Length (m)	Top diameter (cm)
End posts	2 or 2.3	10-13
Struts	2	8-10
Stakes	1.7	5-8

Rabbit and stock	Length (m)	Top diameter (cm)
End posts	2.3	10-13
Struts	2	8-10
Stakes	1.7	8-10

Table 2. Woodwork sizes – sheep and cattle

Sheep	Length (m)	Top diameter (cm)
End posts	2.3	10-13
Struts	2	8-10
Stakes	1.7	8-10

Cattle. Mines and quarries	Length (m)	Top diameter (cm)
End posts	2.3	10-13
Struts	2	8-10
Stakes	1.8	8-10

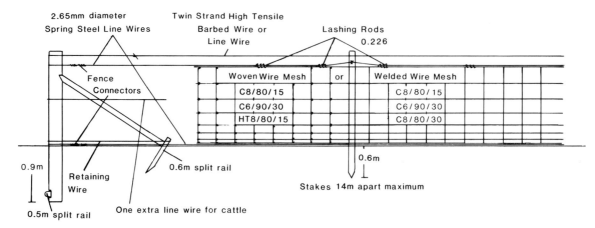

Not to scale. For descriptions of recommended netting and possible alternatives see Figure 18 page 18. Netting sizes and patterns.

Figure 3. Stock fence specification.

rity fencing around quarries and disused mines to meet the requirements of Section 151 of the *Mines and Quarries Act 1954*.

Deer fence – light specification

This basic specification (Figure 5) is only suitable for excluding roe deer or a combination of roe with rabbits or sheep. The light netting

Table 3. Woodwork sizes – roe deer

Roe deer	Length (m)	Top diameter (cm)
End posts	2.8	10-13
Struts	2.5	8-10
Stakes	2.5	5-8

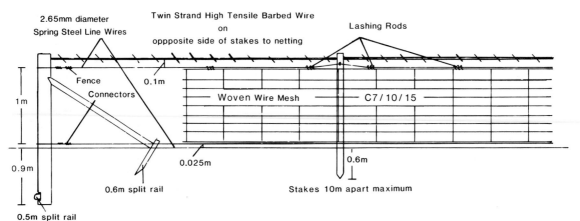

Not to scale. For descriptions of recommended netting and possible alternatives see Figure 18 page 18. Netting sizes and patterns.

Figure 4. Mines and quarries fence specification.

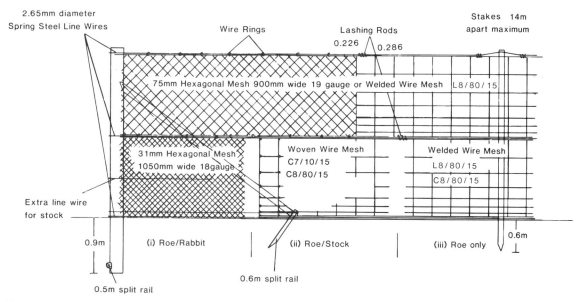

Not to scale. For descriptions of recommended netting and possible alternatives see Figure 18 page 18. Netting sizes and patterns.

Figure 5. Deer fence – light specification for roe.

specified for the upper half of the fence is not robust enough to hold against the larger deer species. However, this netting 1.8 m wide has been used effectively in conjunction with untreated woodwork as a temporary fence to protect restock areas. Woven and welded wire mesh netting with 300 mm spacing between the vertical wires is not recommended because roe deer does and kids can pass through the mesh with relative ease.

Deer fence – heavy specification

This basic specification (Figure 6) should be adopted when fencing out fallow, red or sika deer or any combination of these with roe deer, domestic stock or rabbits.

In upland regions the minimum specification as shown must be increased to compensate for the combined effect of severe weather, difficult soil conditions and heavy deer pressure. Straining post spacing is reduced to 250 – 300 m apart to minimise maintenance by confining any damage in the fence to relatively short strain lengths. Straining post length is increased to 3.2 m to give the greater depth in the ground necessary on all but the stoniest, most compacted soils. A maximum stake spacing of 5 m is required, and an increase in stake length to 2.6 m (2.75 m for peat areas). Stakes may need to be further reinforced with struts (Figure 6) on lengths of fence either exposed to extreme pressure from animals and weather or over shallow soils. The number of gates should be kept to a minimum. They must be of strong construction to the height of the fence, hung on reversed hinges and kept closed with a chain and padlock. Stiles should be provided to reduce the use of gates and therefore the risk of them being left open. Special attention must be paid to filling-in ditches and hollows with additional

Table 4. Woodwork sizes – red, sika or fallow deer

Red, sika or fallow deer	Length (m)	Top diameter (cm)
End posts	2.8	12-18
Struts	2.5	10-13
Stakes	2.6	8-10

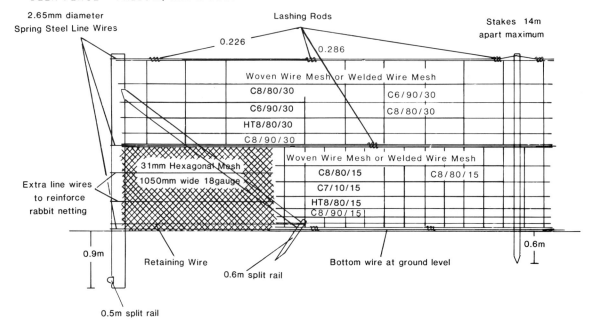

Not to scale. For descriptions of recommended netting and possible alternatives see Figure 18 page 18. Netting sizes and patterns.

Figure 6. Deer fence – heavy specification for fallow, red and sika.

stakes, netting and ground anchors because deer will endeavour to gain access under, rather than through or over the fence.

Deer jumps should be incorporated into the fence to provide an exit for animals that have broken into the plantation. Natural features make the best jumps. The fence is erected along the face of a cutting 1.9 m deep excavated into a slight rise or hillock so that the ramps of the leap run along the line of the fence.

Rivers, burns or streams should be crossed where there is a hard bottom and watergates of substantial construction installed where varying water levels occur. The two gates shown (Figure 7) give the general design principles and the sizes of materials required will depend on the width and depth of the watercourse to be crossed. Watergates and the short lengths of approach fencing should be separate from the main fence. In areas of exceptionally high red deer pressure heavy grade mild steel or high tensile netting may be necessary.

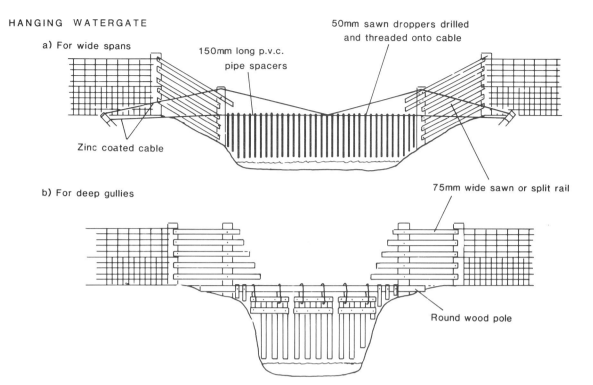

Not to scale. For descriptions of recommended netting and possible alternatives see Figure 18 page 18. Netting sizes and patterns

Figure 7. Watergate specification.

Fence components and associated tools

Metalwork

Before dealing in detail with the different materials and tools it is desirable to understand the meaning of the fencing terms used in this document. Many of these terms have synonyms or sometimes other interpretations which are used in different parts of the country. The accompanying diagram (Figure 8) is intended to remove any confusion.

The adaptation of spring steel wire to forest fencing as outlined in the introduction, necessitated the use of certain special tools. At the same time other improvements in erection techniques utilised new materials and tools. Each component together with the tools required for its erection is described in the following sections (pages 11-24). A complete list of the tools required, together with the suppliers, is to be found in Appendix I (pages 39-40).

Wire and associated tools

Table 5 gives the specification of the most commonly used gauges and sizes of spring steel and mild steel wire. Figure 9 illustrates graphically the differences in behaviour of the two types of wire under various tensions.

During January 1971 the wire manufacturers metricated their agricultural wire diameters. The new metric sizes are not a direct conversion of the old imperial Standard Wire Gauges (Table 5). Spring steel wire, originally only available as an industrial wire, is now marketed as an agricultural wire and therefore complies with the zinc coating requirements of BS 443 (British Standards Institution, 1982).

Care must be taken not to confuse *spring steel wire* with *high tensile wire*. A 3.15 mm diameter *high tensile wire* having a tensile

Table 5. Wire specifications

	Spring steel		Mild steel	
Standard Wire Gauge (SWG)	12	10	10	8
Diameter in millimetres	2.65	3.15	(3.25 3.15	4.10 old SWG sizes) 4.00
Tensile strength (Newtons mm^{-2})	1520	1520	460 – 600	
Breaking strain in Newtons approx.	8000	10 700	3500	5650
Length in metres per 100 kg	2324	1641	1641	1018

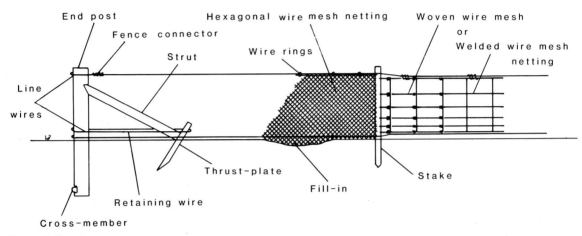

Figure 8. Fencing terms.

strength of 1050 N mm^{-2} is approved for some types of agricultural fences (Ministry of Agriculture, Fisheries and Food, 1969). This quality wire is superior in many ways to mild steel wire but is inferior to spring steel wire particularly with regard to safety. Trials with high tensile wire showed that it had weak spots making it brittle and liable to snap. It is therefore not recommended by the Forestry Commission.

The advantages of *spring steel* over *mild steel fencing wires* can be summarised as follows. The 2.65 mm *spring steel wire* has adequate strength to give a safety margin of 4000 N pull between the top of the 'normal tension' and the breaking point. 'Normal tension' can be defined as the strain that can be obtained by hand operation of the wire strainers without additional levers. The exact amount of tension put on will vary with individual operators, but measurements show that the variation will come within 3100 N to 4000 N, as against the breaking strain of 8000 N. By comparison, *mild steel wires* of 3.15 mm and 4.00 mm have a breaking strain of only 3510 N and 5655 N respectively. In addition, *spring steel wire* has no yield point whereas *mild steel wire* has. The yield point is reached before the wire breaks and is the point at which the wire stretches plastically. Once wire has been stretched plastically it will not return to its original state (Figure 9). Therefore any additional strain applied to the wire after the yield point has been passed will result in the wire stretching, causing it to slacken. Because *spring steel wire* has no yield point and therefore will not stretch and slacken, stakes and straining posts can be spaced at much greater distances than are possible with *mild steel wire*. (The recommended maximum spacing distances are given on pages 6-10.)

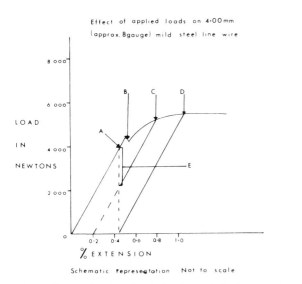
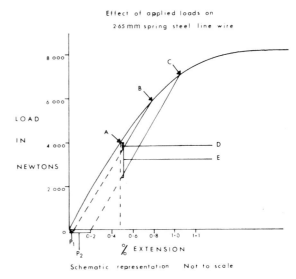

Figure 9. Graphs showing behaviour of steel wires when tensioned.

Left: Mild steel wire.
The wire is initially strained to 4000 N (A). Any external load over the yield point B plastically extends the wire, e.g. to C with no increase in resistance to further deflection. Removal of the load will reduce the original strain by E. Thus extension depends on strain, which, if sufficient e.g. to D, can extend the wire so as to reduce the tension to zero when removed.

Right: Spring steel wire.
The wire is initially strained to 4000 N (A). An external overload of 50% will extend the wire to B. This extension is mainly elastic, but with slight plastic extension P_1. Removal of the load restores the wire to its original line but with a loss in tension of D. If the wire is overloaded by 80% it should extend to C, causing plastic extension P_2. The wire will still restore to its original line, when the load is removed, but with a loss in tension of E.

These two graphs are reproduced by kind permission of Richard Johnson & Nephew (Steel) Ltd.

The tensile strength of spring steel wire is 1490 to 1690 N mm^{-2}. This allows the smaller diameter 2.65 mm wire to be used and still maintain a breaking strain of 8000 N. An advantage of using 2.65 mm wire is that it gives a longer length for a given weight. This reduces transportation costs and as the cost of wire is in part related to weight the price per metre is also less. The spring steel wire is protected against corrosion by a zinc coat of 230 g m^{-2}, which complies with the requirements of BS 443 (British Standards Institution, 1982).

The spring steel wire has great strength for its small diameter but this strength can be drastically reduced if the wire is damaged. Damage can be caused by using unsuitable tools: the grips found on the types of strainers most commonly used on mild steel wire will score spring steel wire. Trewhella Brothers' monkey strainers, illustrated in Figure 10, are the only make so far found suitable. These strainers can be fitted with a spring gauge which indicates the tension on the wire.

Incorrect bending and cutting are other ways of causing damage to the wire. If the wire is to be bent it is essential to make a U-bend with a wire-bending tool (Figure 10) rather than a V-bend. Any wire that is bent to an angle less than three times its diameter will be weakened. The spring steel wire is too hard to cut with conventional pliers. Small hacksaws, wire cutters or Felco C7 cutters are the most satisfactory cutting tools; the latter are illustrated in Figure 10.

One other tool required for handling spring steel wire is a wire dispenser (Figure 11). It is not practical to unwind the coils of wire by hand as it is impossible to prevent kinks developing. A wire dispenser ensures that the wire is laid

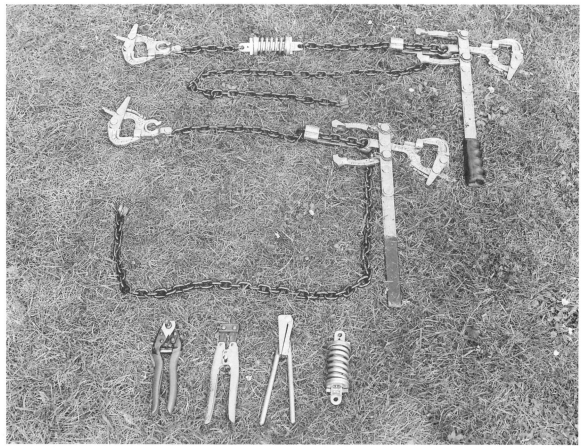

Figure 10. Fencing tools. *Top and middle*: Monkey strainers with and without tension gauge; *bottom, left to right*: Felco C7 wire cutters, 7 in. wire cutters, wire bending tool, and tension gauge. *(B 9417)*

out flat and at the same time eliminates a large proportion of the heavy work. It also speeds up the operation.

Dispensing the wire can be speeded up by using a three-tier wire dispenser (Stickland, 1970). This dispenser enables three wires to be pulled out in one operation. The limitation of the dispenser is that it must be mounted on a vehicle and can therefore only be used when the vehicle can travel along the fence line.

The use of barbed wire on forest fences is not recommended unless neighbouring farmers require it against cattle. Provided the fence is properly erected to the correct specification it will be effective against wild animals and sheep. The addition of barbed wire will not enhance its performance against these animals. The commonly used two-strand mild steel barbed wire (Figure 12) is not suitable for use with spring steel fences where the stakes are spaced 5 m or more apart. The small number of stakes do not give sufficient support and the wire cannot be tensioned enough to prevent sag. A twin strand high tensile barbed wire (Figure 12) is available which is more suitable because it does not sag. It is also preferred to the single strand high tensile barbed wire which is brittle and therefore dangerous if not erected with great care.

Wire fixings

The effectiveness of fences constructed with spring steel wire depends upon the wire retaining its tension. Although spring steel wire does not stretch significantly, the posts may move slightly, causing the wire to slacken. This is particularly likely on short strain distances. In

Figure 11. Wire dispensers with wire. *Left to right*: Donalds, Pulmoflex, and Hayes (Drivall). *(B 9421)*

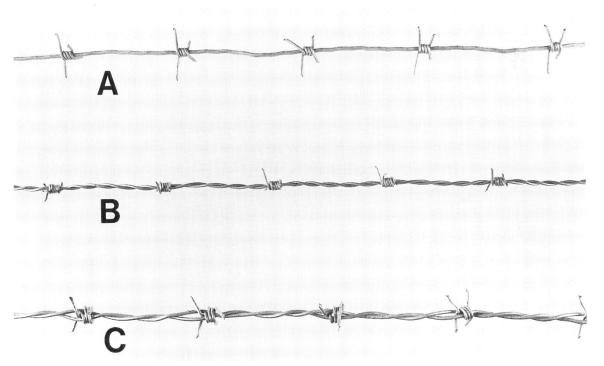

Figure 12. Barbed wire: A– single-strand high tensile barbed wire; B – two-strand high tensile steel barbed wire; C – two-strand mild steel barbed wire. *(A 10770)*

order to restore the fence to full specification some provision to re-tighten the wire easily must be built into the fence.

For this purpose ratchet winders of the type shown (Figure 13) were incorporated in most spring steel fences erected prior to August 1969.

A 9/16 inch A/F square-ended ratchet spanner is required for tightening these.

The ratchet (Figure 13) is now superseded by the cheaper and equally efficient 'Preformed' fence connector. The fence connector (Figure 14 top) is used for both terminating the wire at a

Figure 13. Ratchet winder, now superseded by fence connector shown in Figure 14. *(C 4861)*

post (Figure 35) and for joining the wire (Figure 37) and, if fitted correctly, is stronger than the wire. It is easier to attach than the ratchet, it does not require any tools and, unlike the ratchet, there is no risk of the fence connector damaging the wire.

When it is necessary to join two pieces of spring steel wire, the 'Preformed' fence connector (Figure 14 top) provides a superior join to alternative methods and is simple to fit. The fence connector join is stronger than the wire and therefore perfectly safe. Other designs of wire connector are available (Pepper, 1988) such as the Wirelok, Torpedo, Nicopress sleeves (Figure 15), and the Gripple. The Wirelok may be used as an alternative to the fence connector to secure the retaining wire (Figure 34) and the Wirelok, Torpedo, and Nicopress sleeves may be

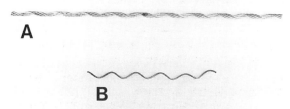

Figure 14. Preformed fixings: A – fence connector; B – lashing rod. *(A 10768)*

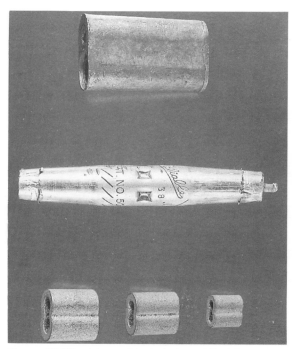

Figure 15. Wire joiners. *Top to bottom:* Wirelok; Torpedo; Nicopress sleeve, which is available in a range of sizes. *(A 10900)*

used to join welded and woven wire mesh. They are particularly useful for joining high tensile woven wire mesh. The Gripple is being evaluated at the time of writing. Knotting the wire, the traditional way of joining mild steel wire, significantly lowers the breaking strain of spring steel and must not be used under any circumstances. The 'double-six' knot (Figure 16) reduces the strength of the wire at least by one third and the 'double-loop' knot by two-thirds.

The correct specifications for use of fence connectors are shown in Table 6.

Zinc coated staples (40 mm × 4.00 mm diameter) should be used for fixing the wire at the appropriate height on the woodwork.

Wire netting and associated tools

There are three types of wire netting; woven wire mesh, welded wire mesh and hexagonal wire mesh. All are manufactured with wire to comply with either BS 4102 (British Standards Institution, 1979) or BS 1485 (British Standards Institution, 1983) and are zinc coated to BS 443 (British Standards Institution, 1982)

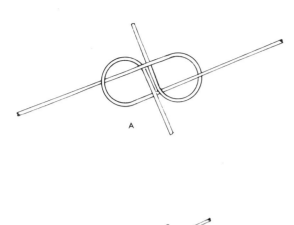

Figure 17. Woven wire mesh netting joints. *Left to right:* hinge joint, tight lock, ring lock.

Figure 16. (A) 'Double-six' knot: *above* – open; *below* – closed. (B) 'Double-loop' knot.

in a variety of gauges and mesh sizes (Figure 18) and are generally sold in 50 m lengths. The exception is high tensile woven wire netting which is available only in 100 m rolls.

Hexagonal netting comes in one basic pattern – as its name suggests.

Woven wire mesh netting is available in the hinge joint and tight lock patterns (Figure 17), UK manufacture of the ring lock pattern having ceased in 1977 although it is manufactured in other countries. Hinged joint netting is manufactured from either high tensile or mild steel wire. This pattern of net has a tendency to concertina vertically when lifted over local humps in the ground and this has to be prevented by using extra stakes. The tight lock joint is only available in high tensile mesh. This joint allows continuous vertical wires which prevent the concertina-ing of the hinge joint.

Welded wire mesh netting is manufactured from mild steel and the horizontal and vertical wires are welded together where they cross. The vertical wires of this type of net are also rigid and therefore prevent concertina-ing. At the

Table 6. Preformed fixings – sizes and uses

Type of fixing	Code	Size (inches)	Use
Fence connector	FC5734802		One for terminating 2.65 mm spring steel wire (Figure 14 top)
			Two for joining together two ends of 2.65 mm spring steel wire (Figures 14 top and 37)
			May be used to join high tensile woven wire mesh netting
	FC5734803		For terminating and joining 3.15 mm spring steel wire
Lashing rods	LR5726011	0.226	For attaching a 2.65 mm spring steel line wire to a C grade woven wire or medium grade welded wire net (Figure 14 bottom)
	LR5726010	0.286	For attaching a 2.65 mm spring steel line wire to two C Grade woven wire or two medium grade welded wire nets. For attaching a 2.65 mm spring steel line wire to a B grade woven wire or heavy grade welded wire net. There is no suitable size of lashing rod available for attaching 3.15 mm line wire to heavy grade netting

time of writing a welded wire mesh with mild steel horizontal wires and high tensile steel vertical wires is being developed. The welded netting is lighter in weight and generally neater in appearance.

Mild steel woven wire and welded wire mesh can be obtained in heavy (B grade) or medium (C grade) quality in a range of sizes (Figure 18). The difference between the heavy grade and the medium grade is in the diameter sizes of wires

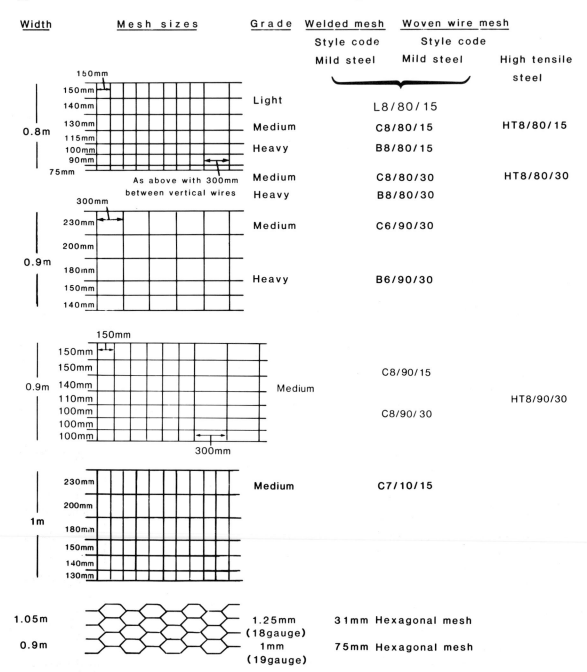

Figure 18. Netting sizes and patterns. Medium grade netting is recommended for most fences: heavy grade netting when increased life is necessary; and high tensile netting where extra strength is required.

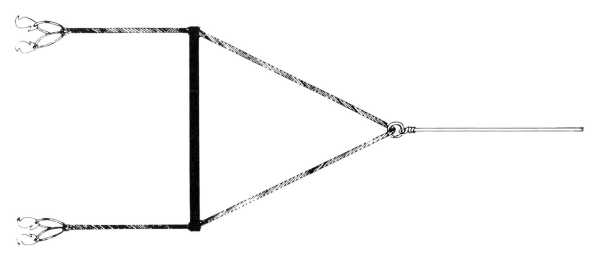

Figure 19. Sharp's straining bar.

used. Heavy grade has the top and bottom horizontal wires of 4.00 mm with all other wires of 3.00 mm. Medium grade has 3.15 mm top and bottom wires with all others of 2.50 mm. Woven wire netting manufactured entirely from 2.50 mm high tensile wire (1050 N mm^{-2}) is now becoming available in an increasing range of mesh sizes. It is lighter in weight than the medium grade mild steel netting and as strong as the heavy grade but the high tensile wire is very hard to bend and tends to be brittle, which makes joining two nets together either a difficult operation or an expensive one if wire joiners such as the Wirelok are used. The medium grade net and high tensile net are considered adequate for most fences. The heavy grade will give an increased life of approximately 30 per cent when compared with the medium grade. In addition to the aforementioned, a light grade welded wire mesh having 2.50 mm and 2.00 mm diameter mild steel wires is available (Figure 18).

All types of woven and welded netting are tensioned, using the wire strainers in conjunction with a straining bar. The design of straining bar can vary (Figures 19, 20, 21 and 22). Hexagonal mesh netting can only be tensioned by hand, otherwise it is pulled out of shape.

The Sharp straining bar (Figure 19) is effective and simple to use. This utilises the clamps of the Monarch strainers and is easily made up in a workshop. Figure 20 shows a straining bar which involves the use of wire rings of approximately 40 mm diameter and a bar. The bar prevents the net 'bagging' in the middle which

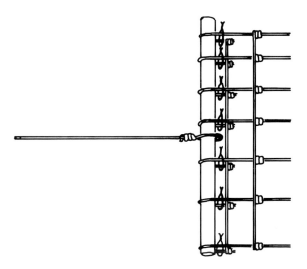

Figure 20. Wire-ring straining bar.

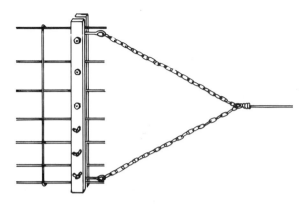

Figure 21. Clamp straining bar.

tends to happen if only the top and bottom wire is pulled as in the Sharp straining bar. Figure 21 shows a straining bar which works on the principle of clamping the wire between two pieces of wood or metal with a series of butterfly or hexagonal nuts and bolts. This method is effective but time-consuming. The 'Tornado' professional clamp (Figure 22) has a patented gripping mechanism which is easy to fit on to the mesh. High tensile netting can only be tensioned successfully using the clamp type straining bars.

Wire netting is attached to the spring steel line wires with either tying wire, pig rings, wire rings, or lashing rods. Tying wire and pig rings are laborious and time-consuming to fix. The wire rings are quickly dispensed from a gun (Figure 23) and are ideal for use with the hexagonal mesh netting. They are not very satisfactory when used with woven wire and welded wire mesh netting as the spring steel line wires and the tensioned netting tend to open these rings. Weight on the net, such as ice or snow, will also open them. A stronger form of attachment, the preformed lashing rod (Figure 14) has been developed. This lashing rod is only intended for attaching welded and woven wire mesh netting to line wires and is manufactured in two sizes for lashing different gauges of wire to different grades of woven and welded wire netting. (The sizes available are shown in Table 6, page 17.)

Wire netting remains the most durable and economic type of netting available despite the introduction of various types of synthetic netting. Both polythene and nylon fibres have been used to make traditional knotted netting for fencing purposes. It has been found that both materials are degraded in sunlight by ultraviolet radiation. Nylon fibres are rotted in less than a year; orange polythene netting will last perhaps twice as long. Fully inhibited black polythene fibres will lose relatively little strength in 6 years and may last up to 10 years. The advantage of polythene netting is that it is light and easily handled. However, experience with the material has shown that animals can easily become entangled in it. For this reason this type of synthetic netting should not be used on any fence, in any circumstances.

A polythene netting similar to the knotted type but with welded joints and stainless steel wire woven into the twine of the horizontal strands is extensively used on farms for temporary electric fencing. To reduce the risk of deer and some other wild animals becoming ensnared in it, it must not be left in position after

Figure 22. The 'Tornado' professional clamp. *(B 9420)*

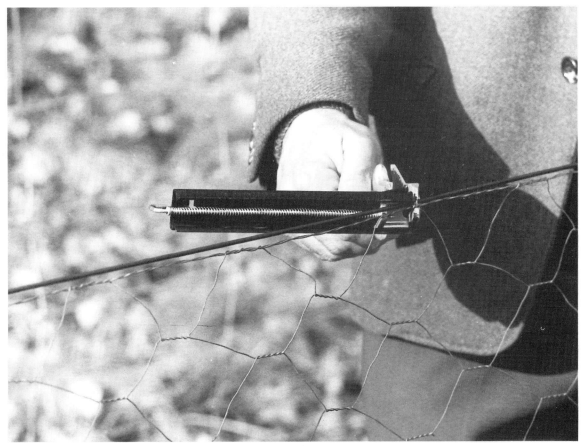

Figure 23. Gerrard wire ring fastener gun in action. *(A 4818)*

the electric fence unit has been turned off. However, even when the unit is on, deer can become entangled in this netting.

New plastics technology has led to the development of a new type of netting. This high tensile polypropylene netting is sufficiently strong to be tensioned in a similar way to the woven-welded wire netting. It may, after further development, prove to be a useful alternative.

Woodwork

Durability, sizes and associated tools

However strong and durable the metal components, the fence will cease to be effective if the woodwork fails. Conversely, it is equally wasteful to erect wooden posts and stakes that will outlive their metal counterparts by many years. Round wood is always preferable to sawn for reasons that are explained later. The strength and durability of all the components in a fence should be matched to the strength of fence required and the period of time for which the fence is needed.

The classification of the natural durability of the heartwood of most timbers is covered in detail in Forest Products Research Laboratory (FPRL) Technical Note No. 40 (1969). Fencing material containing a proportion of sapwood can be classed as perishable and will only have an average life of 5 years. If the fence is required for a period longer than the normal service life of the untreated timber, then it is necessary to treat it with a preservative such as creosote or one of the copper/chrome/arsenic (CCA) formulations (for example 'Celcure A', 'Tanalith C' and

'Treatim CCA'). The timber should be debarked and seasoned to a moisture content of 25% or less before it is treated. A moisture content greater than 25% will inhibit penetration of the preservative.

The preservation of the timber should be either by pressure impregnation or by a full length hot-and-cold open tank treatment with creosote, as described in the FPRL Technical Note No.42 (1969). Different tree species should be treated separately in order to obtain a uniform depth of penetration of preservative. Pine, in particular Scots pine (*Pinus sylvestris*), is a relatively easy species to treat and is to be preferred. Spruce (*Picea* species) and hemlock (*Tsuga heterophylla*) are probably among the most difficult species to treat. Most other softwood species are intermediate between pine and spruce.

The pressure treatment of timber is covered by BS 913, BS 4072 and BS 5589 (British Standards Institution, 1973, 1974 and 1978) which specify minimum retentions of 100 kg m^{-3} for creosote and a minimum dry salt retention of 5.3 kg m^{-3} for CCA preservatives. When treated to these minimum standards pine should have a life of 30 years and spruce 15 to 20 years. Where fencing is required to last much longer than this, 50 years or more, spruce and hemlock should not be used and the timber should be treated to the motorway fencing specifications laid down in BS 913.

There is little difference between the life of creosote-treated and CCA-treated pine because good penetration of preservative can be obtained. This provides a thick protective layer around any untreated wood. In the case of spruce where at best the depth of penetration is relatively shallow, creosote may be superior to CCA preservative. The creosote oil may prevent access of water and fungi to the untreated wood by lessening the timber's tendency to split.

Treated round fencing material lasts longer than sawn material of the same species. The absorbent sapwood provides a protective barrier of treated timber. In general, preservative treated hardwoods do not last as long as similarly treated softwoods.

It is impossible to erect a fence without cutting into some of the treated timber and when doing this it is important to renew the protective layer of preservative. It cannot be over-emphasised that any untreated surface exposed as a result of cutting should be liberally coated with a suitable preservative, for example, an approved organic solvent type wood preservative, creosote, or the concentrated preservative solution, for example 'Celcure B' which meets the requirements of BS 4072 (British Standards Institution, 1974) for cut end treatment. Unfortunately, any retreatment is inferior to the original treatment.

When the timber has been treated with creosote the depth of penetration can easily be seen, but when a copper/chrome/arsenic preservative has been used this is not the case as the preservative is almost colourless. The depth of penetration of this type of preservative can be ascertained by spraying a 1% solution of Chrome Azural S (Mordant Blue 29) on to the cut surface. The solution is red in colour and when it comes into contact with copper of the preservative it turns blue. The extent of the blue colouring clearly defines the penetration of the preservative. The Chrome Azural solution is made up by dissolving 1 g of Chrome Azural in 1000 ml of water containing 2 g of sodium acetate (hydrated).

Any timber treated with CCA preservative must be allowed to stand for 3 weeks before use. After this period the toxic chemicals contained in these preservatives become fixed in the wood and there is no risk to anyone handling the materials. Once the timber has been washed by rain the risk to animals licking the wood will have become negligible. Caution must be exercised in the disposal of wood treated with CCA preservatives. Burning the treated wood in the open should be avoided because there may be a hazard if the smoke is inhaled or if the wood ash is ingested by animals.

Provided an end post is strong enough to withstand the force exerted on it by the strained wire and netting it is wasteful in both material and manpower to erect one of larger diameter. The same can be said about stakes. For example, an end-post with a top diameter of 10 to 13 cm and stakes of 5 to 8 cm top diameter are quite adequate to support a roe deer fence. It will be necessary to increase these sizes by a few centimetres if the fence is to be erected against red deer because this type of fence requires more robust netting, but the principle of using the minimum sizes of timber is the same.

The length of posts, struts and stakes required vary in the first place according to the height of fence to be erected, which is dependent on the species of animal to be excluded, and in the second place, according to the depth the post is to be in the ground. This is dependent on the soil texture.

The utilisation of relatively small diameter posts means that only a small diameter hole need be dug in the ground to accommodate them (Figure 24). The traditional wide hole dug with a spade and often with a step in it is both time consuming and wasteful. The small diameter hole can be dug to the depth required by using the GPO rabbiting tool in conjunction with the 'Shuv-holer' (Figure 25). The GPO rabbiting tool can be modified to act as a rammer to firm the post (Figure 25).

The utilisation of small diameter stakes also makes it possible to use a one-man post driver (Figure 25) to thump them in.

Fencing belt

A leather 'fencing belt' (Figure 26) enables the fencer to have the tools to hand that are most frequently used and most easily lost. A fencing belt should be worn by each member of the fencing team. It has provision for holding a hammer, a pair of Felco wire cutters, a bending tool and a tape. It also has a pouch for staples, and a ring which can be used when pulling out the wire.

Fence construction and maintenance

Construction

When the line of fence has been decided and the fence specification chosen, the amount of material required can be estimated by making a

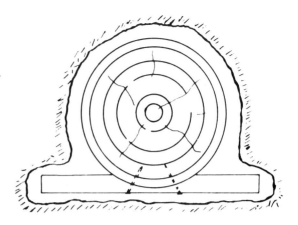

Figure 24. Shape of hole for post. The hole should be just wide enough to accommodate the cross-member and to leave a gap of at least 50 mm around the post.

sketch map while walking the proposed line and marking on the map the locations of straining posts, turning posts, the number of struts re-

Figure 25. Fencing tools: *left to right* – 'Drivall' stake driver; GPO rabbiting tool with rammer modification to handle; 'Shuv-holer' for removing soil from hole. *(B 6686)*

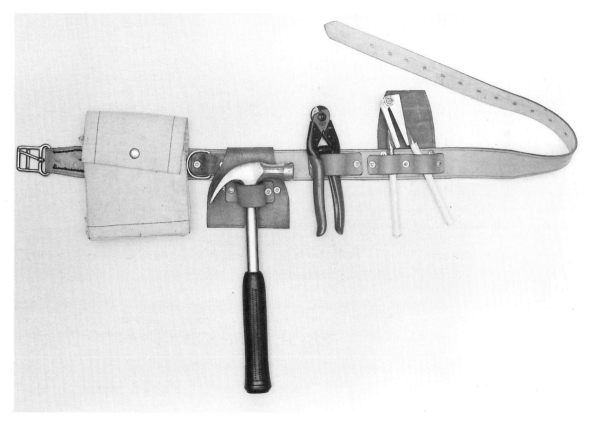

Figure 26. Fencing belt loaded with tools. *(C 4859)*

quired by each post and the distance between stakes (Figure 27).

When it is not possible to distribute the materials along the fence line with a vehicle, carefully sited dumps of materials should be placed within easy reach of the fence line to reduce the distance the material has to be carried by hand. A single dump at the beginning of the fence line should be avoided. If possible the men who are going to erect the fence should be involved in the distribution of the materials.

A two man gang has been found to be the most efficient size of fencing team. If the fence is too long for one team to complete in the time available other fully equipped two man teams can be given sections of the fence. Increasing the size of the gang by one man will not speed up the erection time by one third.

Spacing and erecting the posts

The distance between straining posts is variable and is dependent on the local situation. In theory there is no limit to the distance between straining posts provided they are securely anchored in the ground. In practice the distance is either dictated by the shape and size of the area to be fenced or, where long distances are possible, the spacing for convenience is determined by the length of a coil of 2.65 mm wire (1100 m or 550 m for 50 kg and 25 kg coils respectively).

One of the main advantages of spring steel wire is that it will retain its tension despite fluctuations of temperature. To take advantage of this, great care must be exercised when erecting posts. A poorly erected post will soon be pulled out of the ground, rendering the fence useless.

The holding ability of the post depends upon four factors: the texture of the soil; the depth the post is sunk into the ground; the size of the cross-member; and the amount of soil disturbance round the post.

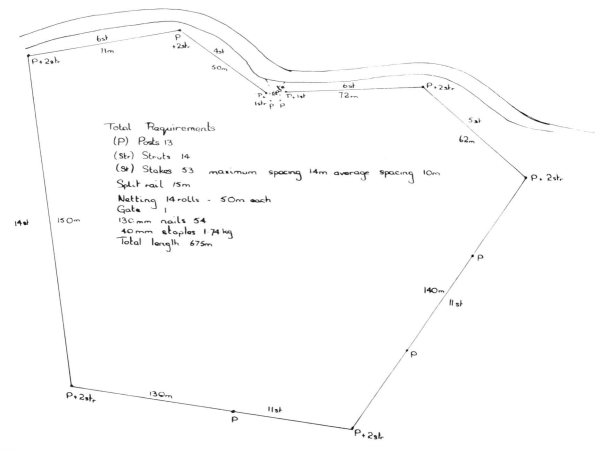

Figure 27. Type of sketch map when estimating quantities of materials required on a fence-line.

The post should be put in to a depth of at least 0.9 m. On areas where the soil type is such that it is considered desirable to go deeper when using mild steel wire for fencing this practice should be continued.

The post hole is dug out with the rabbiting tool and the shuv-holer in the shape shown in Figure 24. This shape of hole produces the minimum soil disturbance around the post while accommodating the cross-member.

The cross-member (Figure 28) is a 0.5 m piece of split rail which is nailed into a notch 75 mm from the bottom of the post. The length of the split rail may be increased for some soil types; for example it may be up to 2 m in peat. However compact the soil type may appear to be, the cross-member should never be omitted as it plays an important part in increasing the stability of the post.

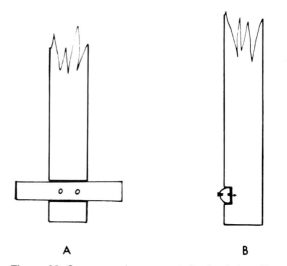

Figure 28. Cross-member on post: A – front view; B – side view.

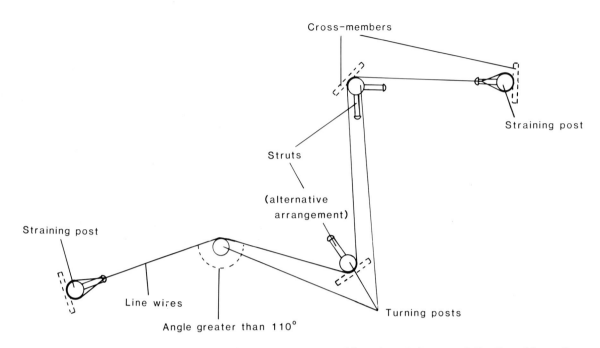

Figure 29. Relative positions of cross-members, struts, posts and line wires at changes of direction of fence-line.

The cross-member should be sited on the opposite side of the post to the line of pull on an end post and opposite the mean line of pull on a corner post (Figure 29). An additional cross-member just below ground level and on the opposite side of the post is sometimes necessary on very wet sites. On areas where it is normal practice to firm the post by placing stones and boulders in the hole this should be continued but not at the expense of omitting the cross-member.

When the post is put into the hole it is leant slightly away from the line of pull. The soil is replaced and rammed tight, particularly the first 0.15 m.

Strutting the posts

The operation of strutting the post makes the assembly into a rigid unit (Figure 30). The strut is located in a notch in the post. The notch is cut to face the direction of strain. In the case of a turning post with only one strut the notch bisects the angle made by the wires. The height of the notch from the ground is dependent on the type of fence being erected but should be ¾ of the distance from the ground to the highest strained line wire (distance H, Figure 30). Therefore for a 0.9 m rabbit or sheep fence the height would be approximately 0.69 m and for a 1.8 m deer fence approximately 1.37 m.

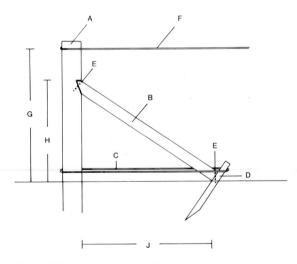

Figure 30. End post assembly: A – post; B – strut; C – retaining wire; D – thrust-plate; E – securing nails; F – highest strained line wire; G – height of line wire; H – height of notch (¾ of G); J – distance from post to base of strut (1¼ × H).

The strut should be long enough to allow the distance (Figure 30 – J) from the base of the strut to the post to be at least 1¼ times the height of the notch from the ground. One end of the strut is shaped to fit the notch in the post. The strut is positioned along the line of the strain or bisecting the angle in the case of a turning post with a single strut (Figure 29).

A thrust-plate is driven into the ground at the base of the strut (Figure 30 – D). The thrust-plate can be either a 0.6 m round stake or pointed split rail with the flat side of the rail against the flat end of the strut (Figure 30). When a round stake is used as a thrust-plate a notch is cut in the end of the strut to hold the two together (Figure 31). The top of the strut is secured to the post and the bottom to the thrust-plate with 130 mm nails.

The post and strut are clamped in position with a separate loop of spring steel wire, the retaining wire (Figure 30 – C), which is as near the ground as possible. This forms the base of a triangle of great strength capable of withstanding the continual strain of the spring steel line wires. The retaining wire prevents the strut being pushed into the ground and the posts splaying out when the fence wires are strained.

The retaining wire is formed by first fixing one end of the wire to the base of the straining post. This is done by making a U-bend in the end of the wire with a bending tool (Figure 10). The U-bend is then fastened to the post with a 40 mm staple in the crook of the bend. Another 40 mm staple holds down the loose end (Figure 32). The wire is taken from the post around the thrust-plate, where it is loosely stapled to retain the wire in position just above the base of the strut, and back around the base of the post. At this point the strainers are attached to the wire and the chain to the thrust-plate above the staple (Figure 33A).

The wire is tightened making sure that the whole of the loop is equally tight. The wire loop is secured either by stapling firmly to the post or preferably by using a fence connector (Figure 33B). Alternatively the retaining wire can be made with a simple loop of tensioned wire which is secured with a wirelok (Figure 34A and B). Struts without retaining wires are sometimes required to brace deer fence stakes (page 9 and Figure 6) to provide extra support on exposed sites or over shallow soils.

Contour and turning posts

Contour posts are used to hold a fence in a depression or gully. They require a cross-member fitted to the bottom of the post but do not need to be strutted. The fence can be held down with an anchor disc where the depression is insufficiently deep to justify a post. However, discs should not be used in wet acid soils because the wire securing the fence to the anchor can rust through within 3 years.

Driven stakes should not be used to hold a fence down unless they are anchored (Figure 38) because the sustained tension on the spring steel line wires will lift them out of the ground.

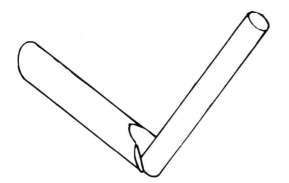

Figure 31. Base of strut notched to a round thrust-plate.

Figure 32. Retaining wire fixed to a post with a U-bend.

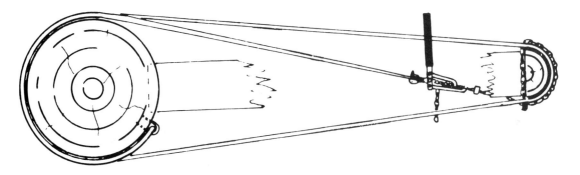

A. The position of the strainers when tightening the retaining wire.

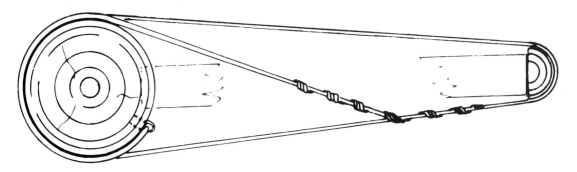

B. Fence connector used to secure the retaining wire.

Figure 33. Straining the retaining wire and securing it with a fence connector.

The function of a stake is to hold a fence upright and not to secure it to the ground.

Turning posts do not require strutting when the fence line changes direction and the internal angle is greater than 110° (Figure 29) but the cross-member may be extended. If the angle is less than 110° the post will require one or two struts (Figure 29). One strut is sufficient provided it does not cause an obstruction to stock or traffic.

Straining the wire

A wire dispenser must always be used when uncoiling wire from a roll. The starting end of each roll of wire is marked with a label and this end is placed on the base of the dispenser so that when the wire is pulled out it comes from the underside of the roll. The clamps on the dispenser should be placed over the roll of wire before the binding wire is cut (Figure 11).

The sequence of operations for straining a wire between two posts A and B will vary slightly according to whether the wire is being pulled out by hand or vehicle. If by hand and the dispenser and wire are at A, the wire is pulled to post B and terminated and the wire is strained at post A. When a vehicle is used the wire is terminated at post A and the wire is dispensed to post B where it is strained.

The wire is terminated at a post with a fence connector by looping the wire around the post, applying half of a fence connector to the wire so that the paint mark is over the end of the wire and applying the remaining half of the fence connector to the wire (Figure 35). The wire loop is held in position at the correct height with a staple driven part way into the back of the post.

The wire is strained at a post by looping the

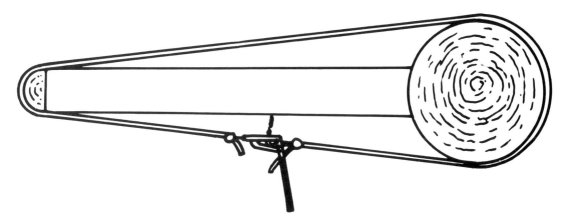

A. The position of the strainers when tightening the retaining wire.

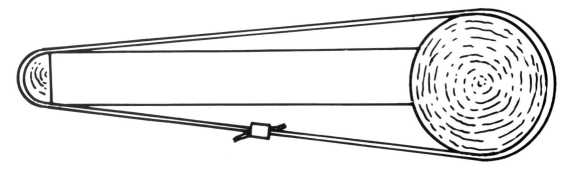

B. Wirelok used to secure the retaining wire.

Figure 34. Straining the retaining wire and securing it with a wirelok.

wire around the post and tightening with the monkey strainers (Figure 36). A staple is part way driven into the post to hold the wire at the required height. The strainers are attached to the wire. When the wire is being tightened on no account should additional levers be added to the strainers. The strainers can be fitted with a tension indicator (Figure 10). An experienced fencer is able to vary the strain he puts on the wire to allow the wire to be raised and lowered

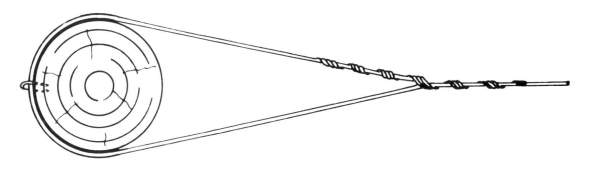

Figure 35. Fence connector terminating a line wire at a post.

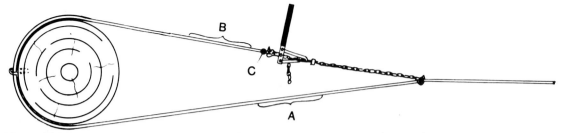

Figure 36. The points on a line wire at which a fence connector is attached (see text).

over undulations in the ground.

When the wire has been tightened it is secured with a fence connector. This is accomplished (Figure 36) by applying half of a fence connector to the wire at A. Points A and B are squeezed together and the remaining half of the fence connector is applied to the wire at B. The strainers are removed, and the wire cut at C.

When cutting the wire it is important to secure either side of the proposed cut to prevent the ends recoiling. This is achieved by placing a foot on one side of the proposed cut and holding the other by hand.

A considerable amount of time can be lost searching for the end of the wire if it has gone back into the coil. When a length of wire has been cut and the coil is not to be re-used immediately it is advisable to put a U-bend in the end to aid recovery. Alternatively if the end is required within a short space of time it can be pushed into the ground.

A part-used roll of spring steel wire should not be removed from the dispenser without first being rebound with tying wire.

Joining the wire

The ultimate strength of the wire is governed by its weakest point. A join in the wire can be a weak point if it is other than a fence connector join (Figure 37). The 'double 6' knot (Figure 16), which was recommended for use before fence connectors were introduced, will reduce the breaking strain of the wire by one third as it will not withstand a pull in excess of 5340 N. The wire joiners (Figure 15) are superior to knots but all make a join that is weaker than the wire. The join made with a Preformed fence connector is always stronger than the wire it has joined.

Two ends of the wire are joined together with two fence connectors. One connector is twisted, half on to one wire and half on the other, so that the wire ends are butt joined at the paint mark. The second connector is twisted over the wire ends with the paint marks lined up (Figure 37).

Spacing the stakes

The maximum distance between stakes will depend on the type of fence required. This is defined in the fence specification (pages 6-11). The maximum spacing should be maintained where the ground is even. It is not practical to adhere rigidly to a given spacing of stakes throughout the length of a fence if it traverses broken and uneven ground. The fence would be ridiculously low in places and metres above the ground in others (Figure 38). Therefore the distance between stakes varies to allow for undulations in the ground but is never greater than the maximum defined by the specification. The irregularity in these distances is not visually unpleasant

Figure 37. Two fence connectors used to join a line wire. *(A 10771)*

because the maximum distances between stakes are greater than with post and rail or mild steel line wire fences. The maximum distance may also have to be reduced along certain stretches of a fence that are likely to be subjected to increased pressure from animals or weather. The effectiveness of a fence should not be compromised by increasing the spacing between stakes just to save money.

Spacing of stakes can only be done during fence erection by using the bottom line wire as a guide. The bottom line wire is tensioned. At the proposed location of the first stake the wire is held to the position it would take on the stake. The height of the wire from the ground from this point to the post is observed. If there is a rise or dip in the ground it will show up and the position of the stake can be moved accordingly to enable the minimum required height of fence to be maintained with the most economical use of stakes and netting fill-in. When the location is fixed the stake is driven in with a 'Drivall' (Figure 25) and the bottom wire is stapled in position. This procedure is repeated progressively up the fence line at each proposed stake location.

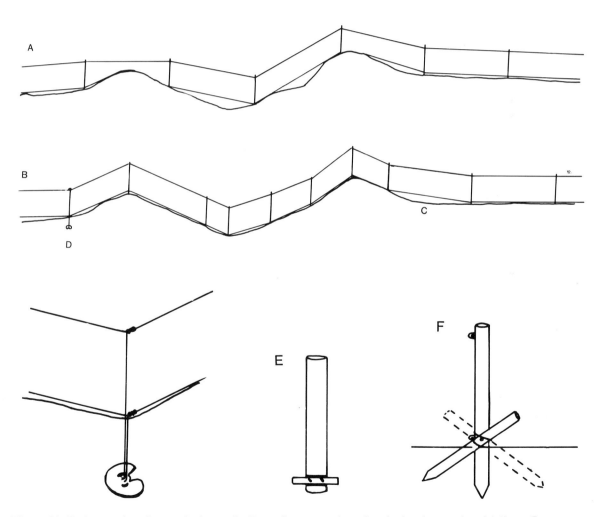

Figure 38. Stake spacing: A – regular intervals; B – stakes spaced to allow for local ground undulations. Gaps under the fence are either: C – filled in with netting or the fence is held down; D – ground anchor; E – with a contour post; F – addition of one or two stakes.

Fixing wire to stakes

Standard 40 mm zinc coated staples are used. Staples are only part way driven into stakes to allow the wire free movement through the staples. This movement is required to enable the wire to be retensioned if necessary. It also allows for movement due to expansion and contraction of the wire caused by temperature changes and reduces the risk of damage to the zinc coating. Staples should be driven at an angle so that their legs are not immediately one above the other. This minimises the risk of the wood splitting and the staple coming out.

Erecting the netting

The erection of hexagonal mesh netting is relatively straightforward. It cannot be strained and all that is required is to pull up the slack by hand. It is joined by overlapping the ends of the two nets and winding the wire ends of each net into the mesh of the other. The netting is attached to the line wires with wire rings (CL22) and stapled to the end posts and stakes. Wire rings zinc coated to BS 443 (British Standards Institution, 1982) should be used (Appendix 1). Preformed lashing rods were designed specifically for use with woven and welded wire mesh netting and are unsuitable for use with hexagonal mesh netting.

Where it is possible to travel the fence line with a vehicle, hexagonal mesh netting can be dispensed a wire netting dispenser (Figure 39). It is also possible for the two-man team to dispense the lightweight 19 gauge 75 mm hexagonal mesh netting by hand using this dispenser. Unfortunately it is not possible to dispense woven wire netting with the dispenser because of the way the netting is rolled by the manufacturer.

Woven and welded wire mesh netting is joined as shown in Figure 40 or with wire connectors (Figure 15). It is tensioned using one of the straining bars, described on pages 19-20. As many as six 50 m rolls of net can be joined together. One end of the net length is stapled to a post and the other end is strained with the straining bar and strainers to the next straining post (Figure 41). Where the distance between posts is greater than 300 m it is necessary to

Figure 39. Wire netting dispenser. *(B 6687)*

strain and attach the netting to a stake. A temporary strut must first be fitted to the stake to enable it to hold the tension on the net until the next length is strained.

When straining high tensile woven wire mesh netting the clamp type straining bar (Figures 21 and 22) is used because all the horizontal wires must be tensioned simultaneously to avoid distorting the mesh.

Before the net is strained the top of the net is clipped to the line wire with two wire rings in each space between stakes. The net is then strained and stapled firmly to a post before the strainers and straining bar are removed. An alternative method is to attach the netting to the posts and strain the netting together using two straining bars (Figure 41). The ends of the netting are joined after straining and before the bars are removed. During the operation of straining the netting, it is sometimes necessary to 'bounce' the net in places to allow it to slide along the line wire. The top and bottom of the netting are finally secured to the line wires with

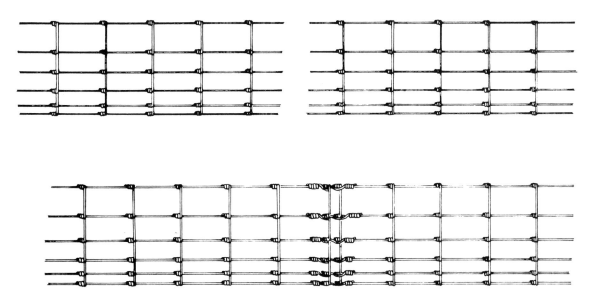

Figure 40. Method of joining woven wire mesh and welded wire mesh netting.

Preformed lashing rods spaced at approximately 2 m centres.

The lashing rod is applied by clipping one end of the rod behind a vertical wire of the net and winding the rod over the horizontal strand of the netting and the line wire. A second lashing rod can be used as a tool to complete the winding of the lashing rod.

Sequence of operations

The method of approach to each operation will vary according to whether the fence erection is being done with the aid of a vehicle or not, but the following basic sequence of operations will remain the same.

Operation one: At least two end posts are assembled with contour and turning posts if required.

Operation two: The centre and bottom wires are pulled out and tensioned.

Operation three: the stakes are arranged at their final spacing, driven into the ground and the bottom wire stapled to them.

Operation four: The remaining wires are run out, tensioned in order from bottom to top and stapled to the stakes.

Operation five: The wire netting is unrolled and joined together. If the netting is woven or welded it is also tensioned and stapled to the straining post. On a deer fence with two layers of netting the lower net is erected first.

Operation six: Hexagonal netting is clipped to the line wires with rings and stapled to the stakes and posts. Woven and welded wire mesh netting is lashed to the line wires with lashing rods and stapled to the stakes and posts.

Operation seven: Any gaps under the netting are filled in or an anchor disc is screwed into the ground to hold the fence down. If rabbit netting is used the bottom 0.15 m is turned out and weighted with sods of earth.

If a fence is erected in sections, the sequence is repeated for each section. It is important that each section is completely finished before starting the next. Any unfilled hollows, gaps or watercourses are quickly exploited as access points by animals which will continue to try to use these places after they are closed. If a vehicle is used the complete fence may be erected as one section in one sequence.

A method of piece-work payment

One of the advantages of spring steel wire fencing is its adaptability to any ground condition. Posts and stakes are not placed at regular dis-

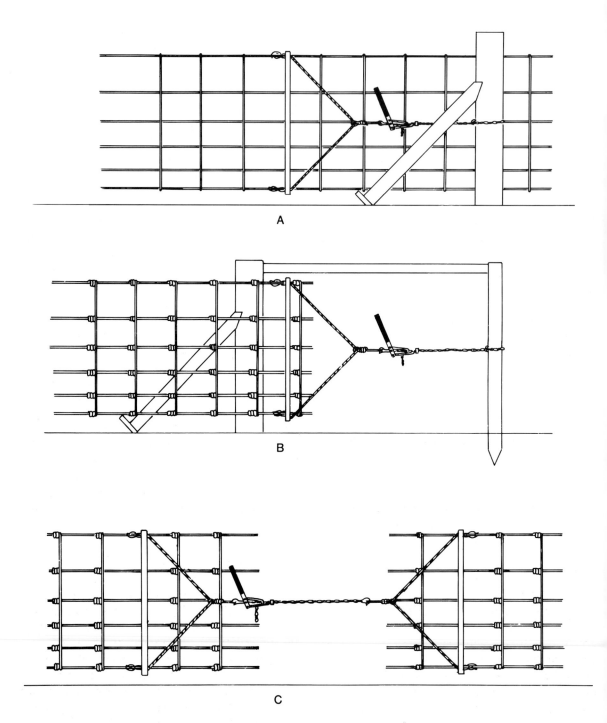

Figure 41. Straining woven or welded wire mesh netting with a straining bar: A – to an end post; B – to an end post using a temporary straining post; C – to the middle.

tances. Within the limits specified on pages 6-11, their positions are determined by the local topography. It is therefore not practical to pay an overall fixed price for a given length (e.g. x pence for every y metres). Each length may have a considerably different work content. A more realistic method of payment is to split the work up into the various operations that are variable between given lengths and to pay a rate for each operation. Some operations such as the erection of line wires have an almost constant work load for the whole length of any fence and this can be paid for by length. The break-down of operations can be summarised as follows.

Erecting the posts. This is the most variable component of the fence as posts can be as far apart as 1000 m. A rate is fixed for the erection of each post. The rate set will depend considerably on the digging conditions (soil type).

Strutting the posts. A post may require one or two struts. (Contour posts and some turning posts do not require struts.) A rate is set for each strut erected.

Stake driving. The number of stakes required depends upon local ground conditions (Figure 38). A price is paid per stake, which could also include an allowance for stapling the line wires to the stake.

Line wire and netting erection and securing. This is a relatively uniform job and a set rate per measured section, e.g. every 20 m, can be paid.

Sodding and filling-in. This can also be paid for on a length basis.

No attempt has been made to suggest what the rates should be in cash as any rate set will be dependent on a wide variety of conditions peculiar to the area being fenced. A work study exercise has been carried out on stock fencing to obtain standard times for each operation. The output guide is available on request from the Work Study Branch, Forestry Commission, Ae Village, Dumfries, DG1 1QB.

Safety

The risk of injury during the erection of a fence utilising 2.65 mm spring steel wire is minimal provided the fencing gang use only the recommended tools and are fully trained in and follow the methods described on pages 23-34 and observe the safety precautions stated in the Forestry Safety Council's leaflet FSC 32 *Fencing*.

The 2.65 mm spring steel wire has a breaking strain of 8000 N. This allows a safety factor of 50%, provided no additional levers are added to the monkey strainers, as the strain applied to the wire will not be greater than 4000 N.

There are two points that should be remembered when handling 2.65 mm spring steel wire. First, that it *is* spring wire so all ends should be held securely to prevent them recoiling and causing injury. Second, that a wire diameter of 2.65 mm is relatively small and any damage that reduces this diameter will weaken the wire considerably.

Tractors and other vehicles must not be used under any circumstances to tension either line wires or wire netting.

Great care must be exercised when dismantling old fences, particularly when vegetation has grown into them. Wire that is pulled can easily and unexpectedly release or break and recoil. It may therefore be necessary to wear additional eye and face protection. Rusty wire and particularly rusty barbed wire will puncture skin very easily. Up-to-date tetanus immunisation is vitally important for all members of the fencing gang.

Maintenance

The amount of maintenance required will largely depend on how well the fence was planned initially. Major repairs should not be required provided the materials were chosen with regard to the anticipated length of life.

A forest fence cannot be economically built to be 100% proof against any animal species so any acceptable specification must be a compromise. It is therefore necessary to make regular checks to close up any entrance holes that may have been made under, between or through the netting. The removal of any beasts may then be desirable to prevent further damage to the fence or plantation. In many cases the risk of break-in can best be reduced by removing the resident animals from an area that has just been fenced rather than by attempting to fence them out.

The soil around a post will inevitably expand and contract due to the influence of changing weather conditions. This movement in some soil types may cause the end post assembly to move slightly which in turn may slacken the line wires. Re-tensioning the wire is a relatively simple task. If the wire is held by a ratchet a few turns with the ratchet spanner will restore the tension. Where the wire is terminated with a Preformed fence connector the wire is re-tensioned with the monkey strainers. The strainers are used to take the strain off the fence connector, half (Figure 42) of which is unwound. The wire is fully tensioned and the fence connector reapplied. Finally the strainers are removed.

Stakes will lift out of the ground if they have been improperly used to hold a fence down in a hollow. In such cases it is insufficient just to re-drive the stakes. The fence must be held down with a contour post or ground anchor before replacing the stakes. Alternatively the gap under the lifted fence can be filled in, using additional netting and stakes. Sticks, stones and turves must not be used as fill-in material because animals can rake it out.

Fences in some areas may be subjected to heavy falls of snow which will flatten them. It would be uneconomical in most cases to build a fence robust enough to withstand the weight of snow. However, when the snow melts, the spring steel wire will spring back bringing the netting with it. The replacement of staples and in some cases stakes will be required to restore the effectiveness of the fence. The fence can be strengthened by putting a strut, at right angles to the fence, on each stake along an exposed length.

Inspections after snowstorms, particularly of new fences, will reveal the points regularly subjected to drifts. The height of the fence can be increased at these points to reduce the risk of animals using the drifts to gain access.

Gate maintenance should not be neglected. Adjustments to hinges and latches may be necessary. Gate posts are frequently hit by vehicles and may need to be repositioned and the gate rehung.

Maintenance costs should be no greater on a spring steel wire fence than those incurred with fences of mild steel wire and in most cases should be considerably less provided, of course, that initially the correct fence specification was chosen and due consideration was given to siting the line of the fence.

ACKNOWLEDGEMENTS

The author wishes to thank all the people involved in the development of the various components of the spring steel forest fence and all those who freely gave information on their experiences with fencing in many varied situations throughout the UK. In particular they wish to thank:

The late Mr M. Goodall for his help and advice during the early development of spring

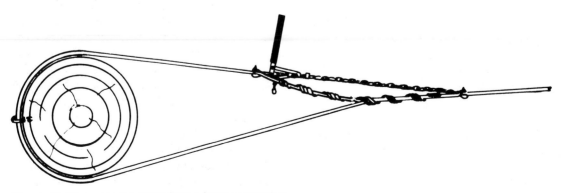

Figure 42. Removing the strain from a fence connector.

steel fencing.

Mr G. Sleigh for his help and much specific information about the properties of spring steel and mild steel wires and Mr J. D. Hoste for providing general technical information on wire and netting.

Mr R. W. Mills for his courage and foresight which resulted in the development and production of welded wire netting.

Mr J. Sutcliffe for his co-operation in the development of 'Preformed' wire fixings for fencing.

Mr D. Mucklow for the provision of new tools and wire fixings for evaluation.

The Training Foresters of the Forestry Commission Education and Training Branch who contributed to the development of the fence erection technique, and to Mr A. Hinde, Conservation Forester, North Scotland Conservancy, for his expert knowledge on red deer fencing in upland conditions.

REFERENCES

BRITISH STANDARDS INSTITUTION (1971). *Pressure creosoting of timber*. BS 913:1971.

BRITISH STANDARDS INSTITUTION (1974). *Wood preservation by means of water-borne copper/chrome/arsenic compositions*. BS 4072:1974.

BRITISH STANDARDS INSTITUTION (1975). *Patented cold drawn carbon steel wire for mechanical springs (Grade 6)*. BS 5216:1975

BRITISH STANDARDS INSTITUTION (1978). *Code of practice for preservation of timber*. BS 5589: 1978.

BRITISH STANDARDS INSTITUTION (1979). *Steel wire for fences*. BS 4102:1979.

BRITISH STANDARDS INSTITUTION (1981). *Battery-operated electric fence controllers not suitable for connection to the supply mains*. BS 6167:1981.

BRITISH STANDARDS INSTITUTION (1982). *Specification for testing zinc coatings on steel wire and for quality requirements*. BS 443:1982.

BRITISH STANDARDS INSTITUTION (1983). *Zinc coated hexagonal steel wire netting*. BS 1485: 1983.

BRITISH STANDARDS INSTITUTION (1987). *Quality systems*. BS 5730: 1987.

BRITISH STANDARDS INSTITUTION (1989). *Fences. Part 2. Specification for rectangular wire mesh and hexagonal wire netting fences*. BS 1722: Part 2: 1989.

DEPARTMENT OF AGRICULTURE AND FISHERIES FOR SCOTLAND AND MINISTRY OF AGRICULTURE, FISHERIES AND FOOD (1977). *Farm capital grant scheme 1973*. Leaflet FCG 1.

FOREST PRODUCTS RESEARCH LABORATORY (1969). *The natural durability classification of timber*. Technical Note 40.

FOREST PRODUCTS RESEARCH LABORATORY (1969). *The hot-and-cold open tank process of impregnating timber*. Technical Note 42.

FORESTRY COMMISSION. *Standard time tables and output guides*. Forestry Commission, Edinburgh.

FORESTRY SAFETY COUNCIL. *Fencing*. FSC 32. Forestry Safety Council, Edinburgh.

McKILLOP, G., PEPPER, H.W. and WILSON, C.T. (1988). Improved specifications for rabbit fencing for tree protection. *Forestry* **61** (4), 359-368.

MELVILLE, R.C., TEE, L.A. and RENNOLLS, K. (1983). *Assessment of wildlife damage in forests*. Forestry Commission Leaflet 82. HMSO, London.

MINISTRY OF AGRICULTURE, FISHERIES AND FOOD (1969). *Permanent farm fences*. FEF Leaflet 6. HMSO, London.

PEPPER, H.W. (1978). *Chemical repellants*. Forestry Commission Leaflet 73. HMSO, London.

PEPPER, H.W. (1988). For easier joining. *Forestry and British Timber*, June, 25.

PEPPER, H.W., CHADWICK, A.H. and BUTT, R. (1992). *Electric fencing against deer*. Forestry Commission Research Information Note 206. Forestry Commission, Edinburgh.

PEPPER, H.W., ROWE, J.J. and TEE, L.A. (1984). *Individual tree protection*. Forestry Commission Arboricultural Leaflet 10. HMSO, London.

POTTER, M.J. (1991). *Treeshelters*. Forestry Commission Handbook 7. HMSO, London.

RATCLIFFE, P.R. and PEPPER, H.W. (1987). *The impact of roe deer, rabbits and grey* squirrels on the management of broadleaved woodlands. Occasional Paper 34. Oxford Forestry Institute.

ROE, M. and TEE, L.A. (1980). Electric and mesh. *Forestry and British Timber*, May, 24-27.

ROWE, J.J. (1976). *Badger gates*. Forestry Commission Leaflet 68. HMSO, London.

STICKLAND, R.E. (1970). Three-tier wire and netting dispenser. *Journal of Agricultural Engineering Research* **15** (3), 314-316.

Appendix 1

Manufacturers' addresses, products and specifications

Product	Manufacturer (see key list)
CCA preservative indicator } Chrome Azural S 'Gurr' } Sodium acetate (hydrated) }	10

Fencing materials – metalwork

Product	Manufacturer (see key list)
Anchor discs (Figure 38)	1
Barbed wire twin strand (Figure 12):	
mild steel	3, 4, 5, 9, 14, 16, 17
high tensile steel	3, 4, 5, 9, 14, 16, 17
Hexagonal mesh netting (Figure 18)	3, 4, 9, 14, 16, 17
Line wire:	
spring steel	3, 5, 9, 14, 16, 17
mild steel	3, 5, 9, 14, 16, 17
Nails, bright or zinc coated	5, 16
Nicopress tool and sleeves (Figure 15)	6
Preformed line fixings (Figure 14)	5, 6, 11, 14
Staples:	
40 mm × 4.00 mm (120 per kg)	3, 4, 5, 9, 14, 16, 17
30 mm × 3.55 mm (220 per kg)	3, 4, 5, 14, 16, 17
Torpedo wire grips (Figure 15)	6
Welded wire mesh (Figure 18)	3, 5, 9, 16
Wire fencing rings CL 22 (Figure 23)	13, 18
Wirelok connectors (Figure 15)	6
Woven wire mesh (Figure 18):	
mild steel	3, 4, 5, 9, 14, 16, 17
high tensile steel	3, 4, 5, 9, 14, 16

Fencing tools

Product	Manufacturer (see key list)
Anchor disc tool	1
Barbed wire cutters	6
Drivall post driver (Figure 25)	4, 6, 9
Felco C7 cutters (Figure 10)	2, 6, 9, 14
(wire cutter 7", wire cutter 12", Maun wire cutting pliers are alternatives)	6, 14
Fencing belt (Figure 26)	6, 7
Fencer's claw hammer	6
Gerrard wire ring fastener gun (Figure 23)	5, 6, 13
GPO rabbiting tool (Figure 25)	6
Monkey strainers (Figure 10)	5, 6, 9, 15
Shuv-holer (Figure 25)	5, 6, 9, 14
Straining bar (Figure 20 and 21) (proprietory names include 'Boundary clamp' or stretcher, 'Tornado' professional clamp – Figure 22)	5, 6, 7, 14
Tension indicator (Figure 10)	6, 7
Wire bending tool (Figure 10)	12
Wire dispenser (Figure 11)	6, 12

Manufacturers' key

1. Biddenden Vineyards Ltd., Little Whatmans, Biddenden, Kent.
2. Burton McCall Ltd., Samuel Street, Leicester LE1 1RU.
3. Centrewire Ltd., PO Box 11, Wymondham, Norfolk NR18 0XD.
4. Estate Wire Ltd., Birley Vale Close, Sheffield S12 2DB.
5. Goodman Croggon Ltd., Polmadie Steelworks, Jessie Street, Glasgow G42 0PG.
6. Drivall Ltd., Narrow Lane, Hurst Green, Halesowen, West Midlands B62 9PA.
7. T L Elliott Trading Ltd., Unit 5C, Ashchurch Business Centre, Tewkesbury, Gloucestershire GL20 8HD.
8. Hunter Wilson & Partners Ltd., Rigg, Gretna, Carlisle CA6 5JL.
9. McArthur Group Ltd., Foundry Lane, Bristol, Avon BS5 7UE and Allender Works, Waterside, Kirkintilloch, Glasgow G66 3NF.
10. Merck Ltd., Broom Road, Poole, Dorset BH12 4NN.
11. Preformed Line Products (GB) Ltd., East Portway, Andover, Hampshire SP10 3LH.
12. Pullmaflex UK Ltd., New Road, Ammanford, Dyfed.
13. J & H Rosenheim & Co Ltd., Glenford Works, Rutherglen, Glasgow G73 1RN.
14. Tornado Wire Ltd., Devonshire Road Estate, Millom, Cumbria LA18 4JS.
15. Trewhella Bros. (UK) Ltd., Rolf Street, Smethwick, Warley, West Midlands B66 2BA.
16. Twil Wire Products Ltd., PO Box 119, Sheffield S9 1TY.
17. Uniwire Ltd., Unit 7, Glanyrafen Industrial Estate, Aberystwyth, Dyfed SY23 3JA.
18. Young Black Industrial Stapling Ltd., 25B Techno Trading Estate, Bramblehead, Swindon.

Note: Addresses were checked before going to press but should be verified before placing order.

Appendix 2

Badger gates
by the late Judith J. Rowe
*former Head of Wildlife Branch,
Forestry Commission.*
Reprinted from Forestry Commission Leaflet 68. Crown copyright 1976.

Introduction

Rabbit fencing may again become an essential protection to new planting. Where badgers occur in large numbers and fence-lines, particularly new ones, cut across their runs, especially in the vicinity of setts, holes are likely to be torn in the netting by the badgers. They may also excavate underneath fences if the soil is suitable. It has been found that careful, progressive erection of a badger gate – whenever possible at the time the fence is erected – can prevent physical damage by badgers to the fence.

Method of erection

If badger gates are essential in a fence, the runs must be located as soon as the fence-line is determined. Work on gates at each badger run crossing the fence line should proceed simultaneously. The work is best done over 6 to 8 weeks in early summer when badger activity is high and runs can be relatively easily recognised. The woodwork of the gate should be treated in the same way as the woodwork of the fence of which it is a part, so that both have approximately the same life expectancy.

At the time the netting is put on the fence a gap is cut in it at the point where the run crosses the fence-line. The gap should measure about 200 mm across and 270 mm high; these dimensions provide the small overlaps needed for stapling the netting to the frame. Where spring steel fencing is in use sods should be put under the ground-level line wire across the run so that the wire is at ground level and can be earthed over. For a week nothing else is done but the fence should be checked daily for signs of damage and to see that badgers are using the gap. If damage to the fence does occur away from the gap, painting the bottom 150 mm of the netting with a smelly, shortlived deterrent such as creosote to about 4 m of each side of the run will help to prevent it. No work should be done on the gaps until badgers are using them.

The next stage is to lay the floor: a block of wood 190 × 40 × 75 mm. This block goes in just below soil level and replaces a sod under the ground-level line wire. The line wire should be stapled to the block. Block and line wire should again be earthed over. In the course of the next few days, badgers passing through the gap will wear away the covering soil but should accept the wired block. A second week should be allowed to pass with daily checks if possible that the run and gateway are in use.

If all is well, the drilled uprights complete with the lintel should be driven in either side of the floor at the beginning of the third week and the netting stapled around the uprights and the lintel. This frame should be driven in to provide a gap 270 mm high and 190 mm wide. During the next week the gateway should be observed and the fence line checked to see if further damage is occurring. If damage does occur, the provision of an additional badger gate should be considered.

If use continues, a wooden half-door can be suspended from the top of the frame, swinging freely from nails through the holes drilled previously in the uprights. Provided use continues, a full-sized door, which will weigh about 1.1 kg and measures 180 × 250 × 40 mm, can be hung after another week. The door should have a 5 mm gap at the sides and 10 mm gaps at top and bottom around it to ensure that it will continue to swing freely whatever the weather. The door can consist of a wooden frame of 40 × 40 mm timber covered with wire mesh not more than 30 mm in mesh size.

Other users

Foxes, and even pheasants, have been observed to use badger gates but rabbits appear to find the weight, and need to push, beyond their ability to learn. There is a danger that rabbits will use the gaps during the gate erection period: however, the danger is much greater that they will use gaps and holes created by the badger in the absence of gates. It is advisable that the fence and its badger gates are erected in the season prior to planting for this reason: any rabbits that do enter or are already established on the enclosed area can then be removed before the trees are at risk.

Maintenance

Once erected, the gates should need little maintenance but they should be checked regularly, especially in their first autumn, to make sure that they have not been blocked by leaves and twigs.

ACKNOWLEDGEMENTS

The method described here owes its development largely to Forester R. J. King and the many keepers, rangers and naturalists who have been concerned with managing wildlife to prevent damage to human interests without causing undue mortality or interference to any wildlife species.

FURTHER READING

King, R. J. (1964). The badger gate. *Quarterly Journal of Forestry* LVIII(4), 311-319.

Neal, E. (1982). *Badgers in woodlands*, 2nd edition. Forestry Commission Forest Record 103. HMSO, London.